SIPMO 2019

SIPMO 2019

Bari, Italy
17–19 October 2019

Congress President
Lorenzo Lo Muzio

Scientific Committee
Gianfranco Favia
Massimo Petruzzi

Special Issue Editors
Giovanni Lodi
Maddalena Manfredi
Alberta Lucchese
Dorina Lauritano
Giuseppe Colella
Guido Lorenzini

MDPI • Basel • Beijing • Wuhan • Barcelona • Belgrade

Special Issue Editors

Giovanni Lodi
University of Milan
Italy

Maddalena Manfredi
University of Parma
Italy

Alberta Lucchese
University of Campania
"L. Vanvitelli"
Italy

Dorina Lauritano
University of Milan-Bicocca
Italy

Giuseppe Colella
University of Campania
"L. Vanvitelli"
Italy

Guido Lorenzini
University of Siena
Italy

Editorial Office
MDPI
St. Alban-Anlage 66
4052 Basel, Switzerland

This is a reprint of articles from the Special Issue published online in the open access journal *Proceedings* (ISSN 2504-3900) in 2019 (available at: https://www.mdpi.com/2504-3900/35/1).

For citation purposes, cite each article independently as indicated on the article page online and as indicated below:

LastName, A.A.; LastName, B.B.; LastName, C.C. Article Title. *Journal Name* **Year**, *Article Number*, Page Range.

ISBN 978-3-03928-108-4 (Pbk)
ISBN 978-3-03928-109-1 (PDF)

Cover image courtesy of Massimo Petruzzi.

© 2020 by the authors. Articles in this book are Open Access and distributed under the Creative Commons Attribution (CC BY) license, which allows users to download, copy and build upon published articles, as long as the author and publisher are properly credited, which ensures maximum dissemination and a wider impact of our publications.

The book as a whole is distributed by MDPI under the terms and conditions of the Creative Commons license CC BY-NC-ND.

Contents

Preface to "SIPMO 2019" xiii

Isaäc van der Waal
Diseases of the Tongue; Some Unusual Lesions and Disorders
Reprinted from: *Proceedings* **2019**, *35*, 1, doi:10.3390/proceedings2019035001 1

Paola Carcieri
Desquamative Gingivitis: Protocols, Procedures and Critical Issues
Reprinted from: *Proceedings* **2019**, *35*, 2, doi:10.3390/proceedings2019035002 4

Olga Di Fede and Giuseppina Campisi
Challenges and Controversies in the Diagnosis of MRONJ
Reprinted from: *Proceedings* **2019**, *35*, 3, doi:10.3390/proceedings2019035003 7

Vito Carlo Alberto Caponio, Carlo Lajolo, Giuseppe Troiano, Claudia Arena, Lorenzo Lo Muzio and Michele Giuliani
Ceaseless Oral Ulcerative Manifestations
Reprinted from: *Proceedings* **2019**, *35*, 4, doi:10.3390/proceedings2019035004 9

Dorina Lauritano, Alberta Lucchese and Massimo Petruzzi
An Unusual Case of Oro-Facial Chronic Pain
Reprinted from: *Proceedings* **2019**, *35*, 5, doi:10.3390/proceedings2019035005 11

Rodolfo Mauceri, Giuseppina Campisi, Claudia Colomba, Olga Di Fede and Vera Panzarella
Mucocutaneous Leishmaniasis Mimicking Squamous Cell Carcinoma
Reprinted from: *Proceedings* **2019**, *35*, 6, doi:10.3390/proceedings2019035006 14

Massimo Petruzzi, Fedora Della Vella, Pasquale Sportelli and Eugenio Maiorano
Unusual Salivary Gland Tumor of the Palate: Clinical, Histological and Immunohistochemical Features
Reprinted from: *Proceedings* **2019**, *35*, 7, doi:10.3390/proceedings2019035007 17

Umberto Romeo, Federica Rocchetti and Alessandra Montori
Criticisms and Controversies in the Diagnosis of Cheilitis
Reprinted from: *Proceedings* **2019**, *35*, 8, doi:10.3390/proceedings2019035008 19

Serban Radu Tovaru
Oral leukoplakia. A More Challenging Disorder than It Seems
Reprinted from: *Proceedings* **2019**, *35*, 9, doi:10.3390/proceedings2019035009 23

Matteo Val, Melania Lupatelli, Marco Ardore, Roberto Marino and Monica Pentenero
Persistent Oral Erosion: A Diagnostic Challange
Reprinted from: *Proceedings* **2019**, *35*, 10, doi:10.3390/proceedings2019035010 25

Davide B Gissi, Andrea Gabusi and Lucio Montebugnoli
A Case of Intra-Oral Bone Exposure of the Hard Palate: A Clinical Diagnostic Dilemma
Reprinted from: *Proceedings* **2019**, *35*, 11, doi:10.3390/proceedings2019035011 27

Claudia Arena, Marco Vairano, Marco Mascitti, Andrea Santarelli, Mario Dioguardi and Khrystyna Zhurakivska
Evaluation of ¡em¿Echinophora tenuifolia¡/em¿ L. Extracts on HSC-2 Cell Line
Reprinted from: *Proceedings* **2019**, *35*, 12, doi:10.3390/proceedings2019035012 29

Dorina Lauritano, Giulia Moreo, Francesco Carinci, Alberta Lucchese, Dario Di Stasio, Fedora Della Vella and Massimo Petruzzi
Preventing Bacterial Leakage in Implant-Abutment Connection: A Review
Reprinted from: *Proceedings* **2019**, *35*, 13, doi:10.3390/proceedings2019035013 30

Dorina Lauritano, Giulia Moreo, Francesco Carinci, Alberta Lucchese, Dario di Stasio, Fedora della Vella and Massimo Petruzzi
Effects of Periodontal Therapy on the Management of Cardiovascular Disease
Reprinted from: *Proceedings* **2019**, *35*, 14, doi:10.3390/proceedings2019035014 32

Dorina Lauritano, Giulia Moreo, Francesco Carinci, Alberta Lucchese, Dario di Stasio, Fedora della Vella and Massimo Petruzzi
Two-Way Relationship between Diabetes and Periodontal Disease: A Reality or a Paradigm?
Reprinted from: *Proceedings* **2019**, *35*, 15, doi:10.3390/proceedings2019035015 33

Dorina Lauritano, Giulia Moreo, Francesco Carinci, Alberta Lucchese, Dario di Stasio, Fedora della Vella and Massimo Petruzzi
Helicobacter Pylory and Oral Diseases
Reprinted from: *Proceedings* **2019**, *35*, 16, doi:10.3390/proceedings2019035016 34

Dorina Lauritano, Giulia Moreo, Francesco Carinci, Alberta Lucchese, Dario di Stasio, Fedora della Vella and Massimo Petruzzi
The Effect of Tobacco Smoking on Periodontal Health
Reprinted from: *Proceedings* **2019**, *35*, 17, doi:10.3390/proceedings2019035017 35

Vito Carlo Alberto Caponio, Giuseppe Troiano, Marco Mascitti, Andrea Santarelli, Rodolfo Mauceri and Lorenzo Lo Muzio
Predicting Death in Patients with Squamous Cell Carcinoma of the Tongue
Reprinted from: *Proceedings* **2019**, *35*, 18, doi:10.3390/proceedings2019035018 36

Mario Carbone, Veronica Avolio, Marco Cabras, Paola Carcieri, Paolo Giacomo Arduino and Roberto Broccoletti
Role of Periodontal Therapy Plus Sodium Doxycycline in the Management of Desquamative Gingivitis: A Pilot Study
Reprinted from: *Proceedings* **2019**, *35*, 19, doi:10.3390/proceedings2019035019 37

Fabio Croveri, Vittorio Maurino, Alessandro d'Aiuto, Marta Dani, Andrea Boggio and Lorenzo Azzi
Primordial Odontogenic Tumour: A Systematic Review
Reprinted from: *Proceedings* **2019**, *35*, 20, doi:10.3390/proceedings2019035020 40

Fedora Della Vella, Sara Galleggiante, Claudia Laudadio, Maria Contaldo, Dario Di Stasio, Marilina Tampoia and Massimo Petruzzi
Serum and Salivary BP180 NC 16a Enzyme-Linked Imunosorbent Assay in Mucous Membrane Pemphigoid. Analysis of a Cohort of 25 Patients
Reprinted from: *Proceedings* **2019**, *35*, 21, doi:10.3390/proceedings2019035021 44

Rodolfo Mauceri, Anna Di Grigoli, Michele Giuliani, Marco Mascitti, Carmine Del Gaizo and Olga Di Fede
Application of Ozone Therapy in the Conservative Surgical Treatment of Osteonecrosis of the Jaw: Preliminary Results
Reprinted from: *Proceedings* **2019**, *35*, 22, doi:10.3390/proceedings2019035022 47

Paolo Junior Fantozzi, Michael Monopoli and Alessandro Villa
Oral Oncology and Oral Medicine Fellowship for the General Dentist
Reprinted from: *Proceedings* **2019**, *35*, 23, doi:10.3390/proceedings2019035023 . . . . **50**

Alessio Gambino, Marco Cabras, Adriana Cafaro, Paolo Giacomo Arduino, Paola Carcieri, Davide Conrotto, Mario Carbone, Luigi Chiusa, Adam Strange, Colin Hopper and Roberto Broccoletti
Use of Optical Coherence Tomography in a Patient with Erosive Oral Lichen Planus Treated with Low-Level Laser Therapy. Preliminary Findings
Reprinted from: *Proceedings* **2019**, *35*, 24, doi:10.3390/proceedings2019035024 . . . . **52**

Gioele Gioco, Romeo Patini, Giuseppe Troiano, Alessia Di Petrillo, Patrizia Gallenzi, Luisa Limongelli and Carlo Lajolo
Oral Manifestations of Psoriasis: A Systematic Review
Reprinted from: *Proceedings* **2019**, *35*, 25, doi:10.3390/proceedings2019035025 . . . . **54**

Gioele Gioco, Cosimo Rupe, Giuseppe Troiano, Michele Giuliani, Massimo Petruzzi and Carlo Lajolo
Teeth Extractions in Subjects Undergoing Radiotherapy for Head and Neck Cancers: A Systematic Review on the Clinical Protocols for Preventing Osteoradionecrosis (ORN). Extractions after Radiotherapy (Part 2)
Reprinted from: *Proceedings* **2019**, *35*, 26, doi:10.3390/proceedings2019035026 . . . . **56**

Davide B. Gissi, Umberto Romeo, Gianluca Tenore, Monica Pentenero, Giuseppina Campisi, Rodolfo Mauceri, Giuseppe Colella, Roberto De Luca, Rosario Serpico, Dario Di Stasio and et al.
Clinical Validation of 13-gene DNA Methylation Analysis from Oral Brushing: A Non Invasive Sampling Procedure for Early Detection of Oral Squamous Cell Carcinoma. A Multicentric Study
Reprinted from: *Proceedings* **2019**, *35*, 27, doi:10.3390/proceedings2019035027 . . . . **59**

Francesca Graniero, Leonardo D'Alessandro, Alessandra Montori, Federica Rocchetti, Vito Cantisani, Andrea Cassoni, Gianluca Tenore and Umberto Romeo
Intraoral Ultrasound in the Evaluation of Depth of Invasion in OSCC. Preliminary Results
Reprinted from: *Proceedings* **2019**, *35*, 28, doi:10.3390/proceedings2019035028 . . . . **61**

Elisa Luconi, Lucrezia Togni, Giovanni Giannatempo, Vito Carlo Alberto Caponio, Marco Mascitti and Andrea Santarelli
p63 Expression in Solitary and Syndromic Odontogenic Keratocysts: An Immunohistochemical Study
Reprinted from: *Proceedings* **2019**, *35*, 29, doi:10.3390/proceedings2019035029 . . . . **63**

Sonia Marino, Roberta Gualtierotti, Valeria Caneparo, Marco De Andrea, Marisa Gariglio, Pier Luigi Meroni, Eleonora Bossi and Francesco Spadari
IFI16 and Anti-IFI16 as Novel Biomarkers for Sjoegren's Syndrome: Preliminary Data
Reprinted from: *Proceedings* **2019**, *35*, 30, doi:10.3390/proceedings2019035030 . . . . **65**

Marco Mascitti, Lucrezia Togni, Lorenzo Lo Muzio, Giuseppina Campisi, Federico Mazzoni and Andrea Santarelli
Odontogenic Cysts: A 30-Year Retrospective Clinicopathological Study
Reprinted from: *Proceedings* **2019**, *35*, 31, doi:10.3390/proceedings2019035031 . . . . **67**

Eleonora Morselli, Roberto Rossi, Luca Morandi, Achille Tarsitano, Andrea Gabusi, Linda Sozzi, Stesi Kavaja and Davide B Gissi
13-Gene DNA Methylation Analysis from Oral Brushing: A Non Invasive Diagnostic Tool in the Follow-Up of Patients Surgically Treated for Oral Cancer
Reprinted from: *Proceedings* **2019**, *35*, 32, doi:10.3390/proceedings2019035032 70

Gian Marco Podda, Federica Rocchetti, Daniele Pergolini, Gaspare Palaia, Gianluca Tenore and Umberto Romeo
Circulating Biochemical Molecular Markers (DNA and RNA) in Head and Neck Cancer: A Narrative Review
Reprinted from: *Proceedings* **2019**, *35*, 33, doi:10.3390/proceedings2019035033 72

Roberto Rossi, Achille Tarsitano, Sofia Asioli, Alice Piastra, Andrea Gabusi, Luca Morandi, Laura Luccarini, Laura Felicetti and Davide B. Gissi
Podoplanin Expression and Its Correlation with Perineural Invasion in Oral Squamous Cell Carcinoma
Reprinted from: *Proceedings* **2019**, *35*, 34, doi:10.3390/proceedings2019035034 75

Cosimo Rupe, Francesco Miccichè, Gaetano Paludetti, Patrizia Gallenzi and Carlo Lajolo
Osteoradionecrosis Rate in Patients Undergoing Radiotherapy for Head and Neck Cancer Treatment: A Six Months Follow-Up of a Perspective Clinical Study
Reprinted from: *Proceedings* **2019**, *35*, 35, doi:10.3390/proceedings2019035035 77

Cosimo Rupe, Gioele Gioco, Giuseppe Troiano, Michele Giuliani, Maria Contaldo and Carlo Lajolo
Teeth Extractions in Subjects Undergoing Radiotherapy for Head and Neck Cancers: A Systematic Review on the Clinical Protocols for Preventing Osteoradionecrosis (ORN). Extractions before Radiotherapy (Part 1)
Reprinted from: *Proceedings* **2019**, *35*, 36, doi:10.3390/proceedings2019035036 79

Valentina Russo, Jacopo Lenzi, Roberto Rossi, Luca Morandi, Achille Tarsitano, Andre Gabusi, Chiara Amadasi, Dora Servidio and Davide B. Gissi
Prognostic Role of DNA Methylation Analysis from Oral Brushing in Oral Squamous Cell Carcinoma
Reprinted from: *Proceedings* **2019**, *35*, 37, doi:10.3390/proceedings2019035037 82

Piermichele Saracino, Claudia Arena, Marco Mascitti, Andrea Santarelli, Vera Panzarella and Lucio Lo Russo
TIMELESS in Head and Neck Squamous Cell Carcinoma: A Systematic Review
Reprinted from: *Proceedings* **2019**, *35*, 38, doi:10.3390/proceedings2019035038 84

Giacomo Setti, Gilda Sandri, Elisabetta Tarentini, Lucia Panari, Adele Mucci, Valeria Righi, Marco Meleti, Cristina Magnoni, Ugo Consolo and Pierantonio Bellini
Salivary ¡sup¿1¡/sup¿H-NMR Metabolomics in Primary Sjögren Syndrome. Preliminary Results of a Pilot Case-Control Study
Reprinted from: *Proceedings* **2019**, *35*, 39, doi:10.3390/proceedings2019035039 86

Antonia Sinesi, Giovanna Mosaico, Martina Cont, Savino Cefola, Giovanni Mautarelli and Cinzia Casu
Gum Hypertrophy in Patients in Fixed Orthodontic Therapy Treated with Topical Probiotic Lactobacillus Reuteri: A Pilot Study
Reprinted from: *Proceedings* **2019**, *35*, 40, doi:10.3390/proceedings2019035040 88

Lucrezia Togni, Marco Mascitti, Corrado Rubini, Rodolfo Mauceri, Lorenzo Lo Muzio and Andrea Santarelli
Adenoid Cystic Carcinoma of Salivary Gland: An Immunohistochemical Study
Reprinted from: *Proceedings* 2019, 35, 41, doi:10.3390/proceedings2019035041 90

Giuseppe Troiano, Khrystyna Zhurakivska, Marco Mascitti, Andrea Santarelli, Giuseppina Campisi and Lorenzo Lo Muzio
Development of a Prognostic Model for Tongue Squamous Cell Carcinoma
Reprinted from: *Proceedings* 2019, 35, 42, doi:10.3390/proceedings2019035042 92

Daniela Adamo, Noemi Coppola, Giulio Fortuna, Elena Calabria, Roberto Carbone and Michele D. Mignogna
Persistent Idiopathic Facial Pain Associated with Patent Foramen Ovale with Right- to-Left Shunt and Hyperhomocysteinaemia: When a Symptom Can Save a Life
Reprinted from: *Proceedings* 2019, 35, 43, doi:10.3390/proceedings2019035043 94

Rita Antonelli, Margherita Eleonora Pezzi, Maria Vittoria Viani, Thelma A. Pertinhez, Eleonora Quartieri, Benedetta Ghezzi, Giacomo Setti, Paolo Vescovi and Marco Meleti
Salivary Metabolic Analysis in Healthy Subjects and Perspectives for Patients with Oral Cancer: Pilot Study and Systematic Review
Reprinted from: *Proceedings* 2019, 35, 44, doi:10.3390/proceedings2019035044 96

Sara Attuati, Valeria Martini, Riccardo Bonacina, Umberto Mariani and Andrea Gianatti
Hard Palate Hyperpigmentation Induced by Chloroquine: A Case Report
Reprinted from: *Proceedings* 2019, 35, 45, doi:10.3390/proceedings2019035045 98

Alessandro Antonelli, Fiorella Averta, Federica Diodati, Danila Muraca, Ylenia Brancaccio, Michele Davide Mignogna and Amerigo Giudice
Plasma Cell Mucositis: A Case Report of an Uncommon Benign Disease
Reprinted from: *Proceedings* 2019, 35, 46, doi:10.3390/proceedings2019035046 100

Vera Panzarella, Alessia Bartolone, Domenico Ciavarella, Andrea Santarelli, Olga Di Fede, Rodolfo Mauceri and Giuseppina Campisi
Use of Optical Coherence Tomography in Patients with Desquamative Gingivitis: A Case Series
Reprinted from: *Proceedings* 2019, 35, 47, doi:10.3390/proceedings2019035047 102

Moreno Bosotti, Francesca Boggio, Anna Mascellaro, Margherita Rossi, Massimo Porrini, Ettore del Rosso and Francesco Spadari
Cartilaginous Choristoma of the Lower Lip
Reprinted from: *Proceedings* 2019, 35, 48, doi:10.3390/proceedings2019035048 105

Adriana Cafaro, Marco Cabras, Alessio Gambino, Marco Garrone, Paolo Giacomo Arduino and Roberto Broccoletti
Describing Clinical and Histological Outcome of Oral Cancer Patients with Recurrent Malignant or Premalignant Oral Lesions: A Retrospective Series with a Follow-Up of 15 Years
Reprinted from: *Proceedings* 2019, 35, 49, doi:10.3390/proceedings2019035049 107

Davide Conrotto, Paolo G. Arduino, Roberto Freilone, Paola Carcieri, Alessio Gambino and Roberto Broccoletti
Management of Oral Hydroxyurea-Related Ulcers: A Cases Series
Reprinted from: *Proceedings* 2019, 35, 50, doi:10.3390/proceedings2019035050 110

Antonio Romano, Maria Rosaria Barillari, Carlo Lajolo, Fedora della Vella, Giuseppe Costa, Alberta Lucchese, Rosario Serpico, Francesca Simonelli and Maria Contaldo
Odontostomatological Findings in Heimler Syndrome: A Case Report
Reprinted from: *Proceedings* **2019**, *35*, 51, doi:10.3390/proceedings2019035051 112

Noemi Coppola, Elena Calabria, Giulio Fortuna, Elvira Ruoppo, Marco Caparrotti and Daniela Adamo
Lacosamide in the Treatment of Trigeminal Neuralgia Refractory to Conventional Treatment Due to Severe Leukopenia Induced by Anticonvulsants
Reprinted from: *Proceedings* **2019**, *35*, 52, doi:10.3390/proceedings2019035052 114

Alessandro d'Aiuto, Maria Pellilli, Marta Dani, Fabio Croveri, Andrea Boggio, Vittorio Maurino and Lorenzo Azzi
Methotrexate-induced Plasma Cell Mucositis: A Case Report of a Previous Undescribed Correlation
Reprinted from: *Proceedings* **2019**, *35*, 53, doi:10.3390/proceedings2019035053 117

Leonardo D'Alessandro, Francesca Graniero, Gian Marco Podda, Gaspare Palaia, Gianluca Tenore, Cira Rosaria Tiziana Di Gioia and Umberto Romeo
445 nm Blue Laser in Excisional Biopsy of a Large Lipoma of the Mouth Floor
Reprinted from: *Proceedings* **2019**, *35*, 54, doi:10.3390/proceedings2019035054 119

Marta Dani, Maria Pellilli, Alessandro d'Aiuto, Lucia Tettamanti, Vittorio Maurino and Lorenzo Azzi
Oral Granular Cell Tumour: A Case Report
Reprinted from: *Proceedings* **2019**, *35*, 55, doi:10.3390/proceedings2019035055 121

Raffaella De Falco, Luca Viganò, Maria Giulia Nosotti and Cinzia Casu
Particular Type of Amalgam Tattoo Associated with Rhizotomy in a Patient with Brain Malignant Tumor: A Diagnostic Dilemma
Reprinted from: *Proceedings* **2019**, *35*, 56, doi:10.3390/proceedings2019035056 123

Andrea Gabusi, Camilla Loi, Davide Bartolomeo Gissi, Andrea Spinelli, Antonio Bernardi and Marina Buzzi
Effectiveness of Topical Application of Heterologous Platelet Rich Plasma (PRP) in Oral Mucous Membrane Pemphigoid. A Report of a Case
Reprinted from: *Proceedings* **2019**, *35*, 57, doi:10.3390/proceedings2019035057 126

Di Fede Olga, Giardina Ylenia, Laino Luigi, Mascitti Marco, Melillo Michele, Capra Giuseppina and Panzarella Vera
HPV-DNA Positive/p16 IHC Negative Oral Squamous Cell Carcinoma: A Case Report
Reprinted from: *Proceedings* **2019**, *35*, 58, doi:10.3390/proceedings2019035058 128

Garibaldi Joseph, Grasso Sara, Piazzai Matteo, Merlini Alessandro and Del Buono Caterina
Craniofacial Fibrous Dysplasia: Diagnosis and Treatment Options
Reprinted from: *Proceedings* **2019**, *35*, 59, doi:10.3390/proceedings2019035059 131

Garibaldi Joseph, Grasso Sara, Piazzai Matteo, Merlini Alessandro and Del Buono Caterina
Modified Double-Layered Flap Technique for Closure of an Oroantral Fistula: Surgical Procedure and Case Report
Reprinted from: *Proceedings* **2019**, *35*, 60, doi:10.3390/proceedings2019035060 133

Garibaldi Joseph, Grasso Sara, Piazzai Matteo, Merlini Alessandro and Del Buono Caterina
The Use of Dorsum of Tongue Flap for the Closure of an Oroantral Fistula with no Contiguous Tissue Available to Be Used: Surgical Procedure and Case Report
Reprinted from: *Proceedings* 2019, 35, 61, doi:10.3390/proceedings2019035061 . . . . **135**

Luisa Limongelli, Angela Tempesta, Saverio Capodiferro, Eugenio Maiorano and Gianfranco Favia
Intraoral Salivary Gland Malignancies: Targeted Surgical Therapy Is Guided by Pre-Operative Mini-Invasive Grading
Reprinted from: *Proceedings* 2019, 35, 62, doi:10.3390/proceedings2019035062 . . . . **137**

Melania Lupatelli, Giovanni Agrò, Alessandro Fornari and Monica Pentenero
Oral Leiomyosarcoma or Low-Grade Myofibrosarcoma: Report of a Challenging Differential Diagnosis
Reprinted from: *Proceedings* 2019, 35, 63, doi:10.3390/proceedings2019035063 . . . . **139**

Giovanna Mosaico, Alessio Chirulli, Antonia Sinesi, Luca Viganò and Cinzia Casu
Atypical Gingival Swelling Unrelated to Plaque and Tartar: Diagnostic Difficulty and Conservative Treatment
Reprinted from: *Proceedings* 2019, 35, 64, doi:10.3390/proceedings2019035064 . . . . **141**

Martina Salvatorina Murgia, Germano Orrù, Luca Viganò, Valentino Garau and Cinzia Casu
Labial Lesion with Heterogeneous Aspects in a Patient with Chronic Renal Failure: Diagnostic Difficulties and Literature Review
Reprinted from: *Proceedings* 2019, 35, 65, doi:10.3390/proceedings2019035065 . . . . **144**

Marco Nisi, Rossana Izzetti, Lisa Lardani, Lucia Scarpata, Maria Rita Giuca and Mario Gabriele
Sublingual Lymphangioma Mimicking a Ranula: A Case Report
Reprinted from: *Proceedings* 2019, 35, 66, doi:10.3390/proceedings2019035066 . . . . **146**

Marco Nisi, Rossana Izzetti and Mario Gabriele
Clear Cell Odontogenic Carcinoma of the Mandible: A Case Report
Reprinted from: *Proceedings* 2019, 35, 67, doi:10.3390/proceedings2019035067 . . . . **149**

Matteo Fanuli, Luca Viganò and Cinzia Casu
Photobiostimulation Therapy in Non-Responsive Oral Ulcerative Aftosis: 3 Cases Reports
Reprinted from: *Proceedings* 2019, 35, 68, doi:10.3390/proceedings2019035068 . . . . **151**

Cinzia Casu, Maria Giulia Nosotti, Matteo Fanuli and Luca Viganò
Photodynamic Therapy in Non-Responsive Oral Angular Cheilitis: 4 Case Reports
Reprinted from: *Proceedings* 2019, 35, 69, doi:10.3390/proceedings2019035069 . . . . **154**

Francesca Pavanelli, Roberto Parrulli, Giuseppe Lizio, Roberta Ippolito, Salvatore Emanuele Teresi and Claudio Marchetti
A Case of Difficult Diagnosis: A Squamous Cell Carcinoma with Bone Exposure and Oro-sinus Communication in a Patient Receiving Alendronate
Reprinted from: *Proceedings* 2019, 35, 70, doi:10.3390/proceedings2019035070 . . . . **157**

Margherita Eleonora Pezzi, Rita Antonelli, Maria Vittoria Viani, Emanuela Casali, Thelma A. Pertinhez, Eleonora Quartieri, Paolo Vescovi and Marco Meleti
Characterization of Bacterial Metabolites in Parotid, Submandibular/Sublingual and Whole Saliva of Healthy Subjects
Reprinted from: *Proceedings* 2019, 35, 71, doi:10.3390/proceedings2019035071 . . . . **159**

Raffaella Castagnola, Irene Minciacchi, Cosimo Rupe, Adele Pesce, Maria Contaldo, Nicola Maria Grande, Luca Marigo and Carlo Lajolo
The Outcome of Primary Root Canal Treatment in Post-Irradiated Patients: A Case Series
Reprinted from: *Proceedings* **2019**, *35*, 72, doi:10.3390/proceedings2019035072 . . . . **161**

Antonia Sinesi, Savino Cefola, Salvatore Grieco, Luca Viganò and Cinzia Casu
A Refractory Labial Fissured Cheilitis Treated with Low Level Laser Therapy (L.L.L.T)
Reprinted from: *Proceedings* **2019**, *35*, 73, doi:10.3390/proceedings2019035073 . . . . **163**

Daniela Sorrentino, Sem Decani, Camilla Zenoni and Andrea Sardella
Nevus in the Oral Cavity
Reprinted from: *Proceedings* **2019**, *35*, 74, doi:10.3390/proceedings2019035074 . . . . **165**

Daniela Sorrentino, Niccolò Lombardi, Chiara Battilana, Sem Decani, Dolaji Henin and Vincent Rossi
Treatment of Symptomatic Mandibular Tori: A Case Report
Reprinted from: *Proceedings* **2019**, *35*, 75, doi:10.3390/proceedings2019035075 . . . . **167**

Andrea Spinelli, Davide Bartolomeo Gissi, Roberto Rossi and Andrea Gabusi
A Case of Nivolumab-Associated Oral Lichenoid Lesions
Reprinted from: *Proceedings* **2019**, *35*, 76, doi:10.3390/proceedings2019035076 . . . . **169**

Angela Tempesta, Luisa Limongelli, Saverio Capodiferro, Massimo Corsalini and Gianfranco Favia
Advanced Stages of Medication-Related Osteonecrosis of the Jaw: From Diagnosis to Surgical Treatment and Rehabilitation with Removable Prosthesis
Reprinted from: *Proceedings* **2019**, *35*, 77, doi:10.3390/proceedings2019035077 . . . . **171**

Salvatore Emanuele Teresi, Gerardo Pellegrino, Roberto Parrulli, Agnese Ferri, Francesca Pavanelli, Riccardo Pirrotta and Claudio Marchetti
MRONJ Treatment with Ultrasonic Navigation: A Case Report
Reprinted from: *Proceedings* **2019**, *35*, 78, doi:10.3390/proceedings2019035078 . . . . **173**

Andrea Tesei, Marco Mascitti, Filiberto Mastrangelo, Vera Panzarella, Alessandra Nori and Andrea Santarelli
Ameloblastic Fibrosarcoma: Report of a New Case
Reprinted from: *Proceedings* **2019**, *35*, 79, doi:10.3390/proceedings2019035079 . . . . **176**

Matteo Val, Margherita Gobbo, Marco Rossi, Mirko Ragazzo and Luca Guarda Nardini
Unusual Manifestations of Oral Follicular Lymphoid Hyperplasia Mimicking Oral Lichen Planus
Reprinted from: *Proceedings* **2019**, *35*, 80, doi:10.3390/proceedings2019035080 . . . . **178**

Caterina Buffone, Flavia Biamonte and Amerigo Giudice
Gene Expression Profiles in Surgical Excision Margins Detected by Tissue Auto-Fluorescence (VELscope™) in Oral Potentially Malignant Disorders (OPMDs) and Oral Squamous Cell Carcinoma (OSCC)
Reprinted from: *Proceedings* **2019**, *35*, 81, doi:10.3390/proceedings2019035081 . . . . **180**

Preface to "SIPMO 2019"

The biennial Congress of the Italian Society of Oral Pathology and Medicine (SIPMO) is an International meeting dedicated to the growing diagnostic challenges in the oral pathology and medicine field. The III International and XV National edition will be a chance to discuss clinical conditions which are unusual, rare, or difficult to define. Many consolidated national and international research groups will be involved in the debate and discussion through special guest lecturers, academic dissertations, single clinical case presentations, posters, and degree thesis discussions. The SIPMO Congress took place from the 17th–19th of October 2019 in Bari (Italy), and the enclosed copy of Proceedings is a non-exhaustive collection of abstracts from the SIPMO 2019 contributions.

Gianfranco Favia & Massimo Petruzzi
Congress Scientific Committee

Extended Abstract

Diseases of the Tongue; Some Unusual Lesions and Disorders [†]

Isaäc van der Waal

Department of Oral and Maxillofacial Surgery and Oral Pathology, University Medical Center Amsterdam and Academic Centre for Dentistry Amsterdam (ACTA), P.O. Box 7057, 1007 MB Amsterdam, The Netherlands; i.vanderwaal@hotmail.com

† Presented at the XV National and III International Congress of the Italian Society of Oral Pathology and Medicine (SIPMO), Bari, Italy, 17–19 October 2019.

Published: 10 December 2019

Many lesions and disorders of the oral cavity may affect the tongue as well. On the other hand, some lesions have a strong preference for occurrence on the tongue or may be even limited to the tongue. A selection of both categories will be discussed, emphasizing the diagnostic and management aspects.

Lymphangioma is a developmental disorder that arises at a young age. The clinical presentation is more or less diagnostic. Nevertheless, the taking of a biopsy is recommended. Treatment possibilities are limited, except for small lesions that may be removed without causing much morbidity.

Geographic tongue is not so much a rare lesion, but probably often remains undiagnosed by clinicians, particularly when occurring in children (Figure 1). A rare phenomenon is the occurrence of geographic tonguelike lesion elsewhere in the mouth, being referred to as ectopic geographic tongue or geographic stomatitis. There are no possibilities to cure geographic tongue and the disease may last lifelong.

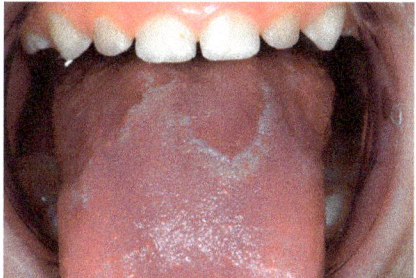

Figure 1. Geographic tongue in a 3-year-old boy.

Median rhomboid glossitis is usually easy to diagnose based on clinical features alone. If a biopsy is taken, it should be realized that the pathologist may be challenged by the presence of elongated rete ridges that may mimic squamous cell carcinoma. Treatment is only indicated in case of symptoms and consists of elimination of possible causative factors, e.g., tobacco habits and the use of corticoid inhalation spray for pulmonary disease and the use of topical antifungals. In rare instances a *Kaposi sarcoma* may arise in the foramen cecum area, somewhat mimicking median rhomboid glossitis. It should be realized that Kaposi sarcoma (KS) may be the first manifestation of an underlying HIV-infection. At the same time, oral KS has been reported in immunocompetent patients [1].

Lingual papillitis is a poorly understood inflammation of the fungiform papillae, showing a quite distinct clinical picture. Lingual papillitis is usually self-healing ('Transient lingual papillitis') [2].

A persistent ulcer on the dorsum of the tongue may have a specific cause, e.g., syphilis I. The occurrence of a squamous cell carcinoma at that particular site is rare. A rather rare entity, mainly occurring on the tongue, is *traumatic ulcerative granuloma with stromal eosinophilia* (TUGSE). The histopathologic features are quite diagnostic. Occasionally, CD30 positive lymphocytes may indicate a peculiar type of T-cell lymphoma. In such cases the patient should be staged for possible involvement elsewhere in the body [3].

A *granular cell tumor* may occur everywhere in the body but has a strong preference for the mouth, particularly for the tongue. The diagnosis is based on histopathologic aspect. A well-known pitfall is the occurrence of pseudoepitheliomatous hyperplasia of the overlying epithelium, that may be misdiagnosed as squamous cell carcinoma.

Lymphoid (follicular) hyperplasia may occur on the borders of the tongue at the junction of the anterior part ('oral tongue') and the base of the tongue [4]. There is usually a bilateral presentation of slightly swollen, soft elastic mucosa. Symptoms are usually absent and in such event a biopsy nor follow-up is indicated. In symptomatic cases, particularly when unilateral, the possibility of a squamous cell carcinoma should be considered.

A range of lesions and conditions may present as bilateral white changes at the borders of the tongue (Table 1). These lesions can sometimes be diagnosed based on the presence of similar lesions elsewhere in the oral cavity, e.g., morsicatio, but others may require a biopsy (Figure 2). This is particularly true when *hairy leukoplakia* is suspected in a patient with a negative medical history. The histopathologic features, including positivity of an Epstein Barr Virus (EBV) immunohistochemical stain, are diagnostic. Although hairy leukoplakia is mainly known as manifestation of an underlying HIV-infection, also other causes of immunosuppression may result in this lesion.

Table 1. Differential diagnosis of bilateral white lesions of the tongue (in alphabetical order).

Candidiasis, hyperplastic
Hairy leukoplakia
Leukoplakia ('true')
Lichen planus
Morsicatio
Pachyonychia congenita
Syphilis, second stage
White sponge nevus

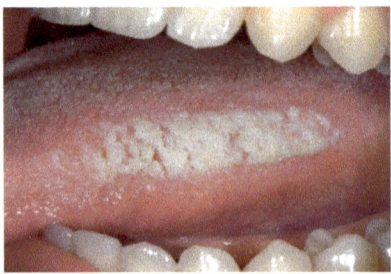

Figure 2. Verrucous lesion, bilateral, on the border of the tongue, being caused by morsicatio.

In the *second stage of syphilis* multiple white lesions ('plaques muceuses') may occur in the oral mucosa, often of the tongue. Another manifestation may be the occurrence of red, patchy and sometimes aphthouslike changes of the oral mucosa, particularly on the dorsal surface of the tongue.

Such lesions may follow a recurrent pattern. A suspected diagnosis of syphilis should be confirmed by serological tests.

A rather unusual tumor, often benign and more or less limited to the anterior tongue, is the *ectomesenchymal chondromyxoid tumor*. The clinical presentation of the tumor is not characteristic, just being a non-ulcerative firm elastic swelling with an intact mucosal surface. Since its first description in the nineties of the last century less than fifty cases have been reported. The challenge is with the histopathologic interpretation, including the use of various immunohistochemical stains. The tumor may be wrongly diagnosed, e.g., as a salivary gland tumor or a (rhabdomyo)sarcoma [5].

Conflicts of Interest: The author declares no conflict of interest.

References

1. Pantanowitz, L.; Khammissa, R.A.; Lemmer, J.; Feller, L. Oral HIV-associated Kaposi sarcoma. *J. Oral Pathol. Med.* **2013**, *42*, 201–207. doi:10.1111/j.1600-0714.2012.01180.x.
2. Kalogirou, E.M.; Tosios, K.I.; Nikitakis, N.G.; Kamperos, G.; Sklavounou, A. Transient lingual papillitis: A retrospective study of 11 cases and review of the literature. *J. Clin. Exp. Dent.* **2017**, *9*, e157–e162. doi:10.4317/jced.53283.
3. Sharma, B.; Koshy, G.; Kapoor, S. Traumatic Ulcerative Granuloma with Stromal Eosinophila: A Case Report and Review of Pathogenesis. *J. Clin. Diagn. Res.* **2016**, *10*, ZD07–ZD09.
4. Stoopler, E.T.; Ojeda, D.; Elmuradi, S.; Sollecito, T.P. Lymphoid Hyperplasia of the Tongue. *J. Emerg. Med.* **2016**, *50*, e155–e156. doi:10.1016/j.jemermed.2015.09.042.
5. Kato, M.G.; Erkul, E.; Brewer, K.S.; Harruff, E.E.; Nguyen, S.A.; Day, T.A. Clinical features of ectomesenchymal chondromyxoid tumors: A systematic review of the literature. *Oral Oncol.* **2017**, *67*, 192–197.

© 2019 by the authors. Licensee MDPI, Basel, Switzerland. This article is an open access article distributed under the terms and conditions of the Creative Commons Attribution (CC BY) license (http://creativecommons.org/licenses/by/4.0/).

Extended Abstract

Desquamative Gingivitis: Protocols, Procedures and Critical Issues †

Paola Carcieri

Department of Surgical Science. Oral Medicine Section, University of Turin, UNITO Lingotto Dental Institute, Via Nizza 230, 10126 Turin, Italy; carcieri.paola@libero.it; Tel.: +39-339-578-4949

† Presented at the XV National and III International Congress of the Italian Society of Oral Pathology and Medicine (SIPMO), Bari, Italy, 17–19 October 2019.

Published: 10 December 2019

1. Introduction

Desquamative gingivitis (DG) isn't a specific disorder; it simply represents the gingival manifestation associated with some heterogeneous mucocutaneous disorders, such as oral lichen planus (OLP), mucous membrane pemphigoid (MMP), pemphigus vulgaris (PV), plasma cell gingivitis (PCG) and few others. Epithelial desquamation, erythema and erosive and/or vesiculo-bullous lesions on the gingiva usually characterise it. The DG appears more frequently in old women and menopause, although it can debut in young people and children. Clinically, it presents moderate pain, partly due to the deposit of plaque in gingival margin, being in some cases the first manifestation of the disease.

Usually of unknown aetiology, the most probable hypothesis is the autoimmune origin of these disease. For these reasons topical treatment has been indicated, as the use of corticosteroid in different forms and prescribed with different posologies, or also systemic drugs administration such as corticosteroids, other immunosuppressants and broad-spectrum antibiotics.

Even if this condition has been reported as non-plaque induced, effective dentogingival plaque control sometimes could resolve the gingival inflammation.

Some studies of periodontal status in patients with DG suggest that in patients MMP the gingivo-periodontal status is worse than health control, in the same case, patients with OLP and PV present deeper pockets and higher loss of the clinical attachment level [1].

A recent study revealed a variation of the "microbiota" in DG-patients with prevalence of some bacterial strains as high-risk pathogen Aggregatibacter Actinomycetemcomitans or moderate risk pathogens Eikenella Corrodens or Fusobacterium Nucleatum [2].

For these reasons, we postulated that efficient plaque control could have been helpful in treating DG.

We report the results of various prospective studies in which the purpose was to evaluate the clinical efficiency of an oral hygiene protocol in patients affected by DG [3–7].

2. Results

From January 2006 to August 2019 at the Oral Medicine Section, University of Turin, we followed 150 patients affected by DG as shown in Figure 1. The patients received a complete periodontal examination at baseline visit, during protocol and at final treatment, including: full mouth plaque scores (FMPS) Sillness and Loe (grading 0-1-2-3), full mouth bleeding upon probing scores (FMBS), probing pocket depth (PPD), mobility index sec Miller (grading 1-2-3), DGCS by Arduino et al. (score 0–4) and detection of perceived pain with VAS scale. All patients received non-surgical periodontal therapy, including oral hygiene instructions, supra- and sub-gingival scaling if required. Oral hygiene instructions were given by the same experienced dental hygienist; she also provided thorough supra-gingival scaling with elimination of all deposits, over three visits and completion within 3 weeks. During each visit, subjects were instructed about proper oral hygiene maintenance at home. Moreover, the instructions were reinforced at each visit and personalised whenever necessary. A cohort of patients was also treated in combination with a solution of sodium iodide associated to salicylic acid (SISA). The solution was used at the end of each session, with an impregnated gauze (with 5 mL of the solution) applied for 15 minutes for both the dental arches, and for other 15 minutes with new gauzes. A second cohort received alternative approaches with ozone therapy. Its use in oral pathology is based on the principle of the liberation of O_3 through the formation of an electromagnetic field, similar to what happens during the discharge of lightning in nature (Chapman effect). The ozone therapy has antimicrobial power (against aerobic and anaerobic bacteria, fungi, viruses), exerts a stimulation of the circulatory system with improvement of tissue oxygenation and modulation of immune cell and shows pain reduction and repeated insufflations stimulate neoangiogenesis with formation of epithelial tissue.

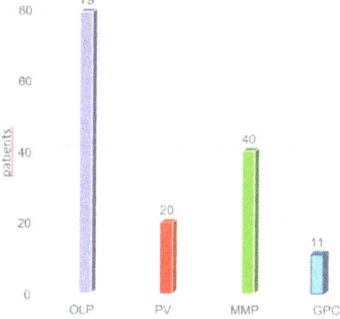

Figure 1. subdivision by type of DG of the 150 patients followed from January 2006 to August 2019 at the Oral Medicine Section, University of Turin.

In our various studies [1,3–6], we have shown that professional oral hygiene procedures are related to significant improvements of gums status, and decreases related pain, in subjects affected by autoimmune diseases as DG with specific gingival localization (Figure 2).

	T0	T4	T5	p*
Full mouth bleeding score (%)	59.00 ± 16.25	-	23.88 ± 11.11	.032
Full mouth plaque score (%)	45.02 ± 13.63	-	29.20 ± 2.73	.038
Probing depth (mm)	2.48 ± 0.51	-	2.29 ± 0.67	.791
Referred symptoms (VAS score)	5 ± 1.32	3.38 ± 1.19	2.44 ± 1.42	.000
Activity score	6.48 ± 1.20	3.08 ± 1.77	2.81 ± 1.79	.005

*Comparative statistics were performed between T0 and T5. Paired samples test was used to test the difference in FMBS, FMPS and PPD. Wilcoxon's signed rank was used to calculate the significance of the patient related outcomes and in gingival clinical outcome (activity score).

Figure 2. results of SISA protocol reported in Carcieri P et al. [6].

3. Conclusions

Regular plaque and tartar removal is therefore a fundamental requirement for a favorable prognosis at a distance in DG-patients [7]. Non-surgical periodontal therapy, and oral hygiene instructions, in combination with S.I.S.A or ozone therapy, could be a possible mean in reducing clinical gingival inflammation and improve patient related outcomes in the different forms of DG. These protocols are conceived to recommend this as complementary line therapeutic intervention, especially in patients with only gingival involvement, during any other medical treatment, helping affected patients in maintaining a good oral hygiene.

Conflicts of Interest: The author declares no conflict of interest. The funding sponsors had no role in the design of the study; in the collection, analyses, or interpretation of data; in the writing of the manuscript, and in the decision to publish the results.

References

1. Arduino, P.G.; Farci, V.; D'Aiuto, F.; Carcieri, P.; Carbone, M.; Tanteri, C.; Gardino, N.; Gandolfo, S.; Carrozzo, M.; Broccoletti, R. Periodontal status in oral mucous membrane pemphigoid: Initial results of a case-control study. *Oral Dis.* **2011**, *17*, 90–94.
2. Arduino, P.G.; Romano, F.; Sasia, D.; Broccoletti, R.; Ricceri, F.; Barbui, A.M.; Brossa, S.; Cipriani, R.; Cricenti, L.; Cabras, M. Subgingival Microbiota in White Patients with Desquamative Gingivitis: A Cross-Sectional Study. *J. Periodontol.* **2017**, *88*, 643–650.
3. Arduino, P.G.; D'Aiuto, F.; Cavallito, C.; Carcieri, P.; Carbone, M.; Conrotto, D.; Defabianis, P.; Broccoletti, R. Professional Oral Hygiene as a Therapeutic Option for Pediatric Patients with Plasma Cell Gingivitis: Preliminary Results of a Prospective Case Series. *J. Periodontol.* **2011**, *82*, 1670–1675.
4. Arduino, P.G.; Lopetuso, E.; Carcieri, P.; Giacometti, S.; Carbone, M.; Tanteri, C.; Broccoletti, R. Professional oral hygiene treatment and detailed oral hygiene instruction in patients affected by mucosa membrane pemphigoid with specific gingival localization in 12 patients. *Int. J. Dent. Hyg.* **2012**, *10*, 138–141.
5. Gambino, A.; Carbone, M.; Arduino, P.G.; Carcieri, P.; Carbone, L.; Broccoletti, R. Conservative approach in patients with pemphigus gingival vulgaris: A pilot study of five cases. *Int. J. Dent.* **2014**, *2014*, 747506, doi:10.1155/2014/747506.
6. Carcieri, P.; Broccoletti, R.; Giacometti, S.; Gambino, A.; Conrotto, D.; Cabras, M.; Arduino, P.G. Favourably effective formulation of sodium iodide and salicylic acid plus professional hygiene in patients affected by desquamative gingivitis. *Biol. Regul. Homeost. Agents* **2016**, *30*, 1141–1145.
7. García-Pola Vallejo, M.J.; Rodriguez-López, S.; Fernánz-Vigil, A.; Bagán Debón, L.; García Martín, J.M. Oral hygiene instructions and professional control as part of the treatment of desquamative gingivitis. *Syst. Med. Oral Patol. Oral Cir. Bucal.* **2019**, *24*, e136–e144.

© 2019 by the authors. Licensee MDPI, Basel, Switzerland. This article is an open access article distributed under the terms and conditions of the Creative Commons Attribution (CC BY) license (http://creativecommons.org/licenses/by/4.0/).

 *proceedings*

Extended Abstract

Challenges and Controversies in the Diagnosis of MRONJ †

Olga Di Fede * and Giuseppina Campisi

Department of Surgical, Oncological and Oral Sciences, University of Palermo, 90133 Palermo, Italy; campisi@odonto.unipa.it

* Correspondence: odifede@odonto.unipa.it
† Presented at the XV National and III International Congress of the Italian Society of Oral Pathology and Medicine (SIPMO), Bari, Italy, 17–19 October 2019.

Published: 11 December 2019

Medication-related osteonecrosis of the jaw (MRONJ) is a relatively rare but potentially serious and debilitating complication. It consists of progressive bone destruction in the maxillofacial area of patients exposed to the treatment with drugs associated with the risk of ONJ (antiresorptive and antiangiogenic agents), in the absence of a previous radiation treatment [1].

In order to adjudicate MRONJ case, many considerations are essential for the clinician.

Firstly, in absence of bone exposure and or fistula, other signs or symptoms (e.g., abscess, periodontal instability, presence of swelling and/or pus, severe pain, lockjaw, sequestrum) are not diagnosed. This is due to restricted criteria of AAOMS definition in which these clinical findings are not included but already recognized and reported in literature [2–4].

Furthermore, many times the clinical history of the patient is not complete and all identified and reported drugs related to ONJ are not identified from the dentist, the practitioner, or the oncologist. Many clinicians take in account only bisphosphonates as drugs related to ONJ risk. Nowadays, many other drugs have been related to this adverse event, from antiresorptive to antiangiogenic agents (e.g., Bisphosphonates, Denosumab, Bevacizumab, Sunitinib). The lack of knowledge and the necessity of a continuous update is an essential element for all clinicians.

After, other critical point for MRONJ diagnosis is the disclaimer of imaging whereas this is essential for unexposed clinical form of osteonecrosis. The adding of radiological findings in case of suspicious of MRONJ determine not also an underestimation of frequency data but also a delay for staging and management [5,6].

It is important to note that in few years, many data are added to the definition of MRONJ and, probably, much more should be discovered. This is the main controversy in terms of diagnosis. The uniformity in terms of clinical and radiological findings must be achieved in short time since the underestimation and delay diagnosis have to avoid in order to support the affected patient, already overloaded from primary disease (i.e., cancer).

The main challenge is to create a multidisciplinary network for a standardized approach, with a sustained dialogue among specialists involved, should be always adopted in order to improve the efficacy of diagnosis process and to ameliorate the patient's quality of life

References

1. Di Fede, O.; Panzarella, V.; Mauceri, R.; Fusco, V.; Bedogni, A.; Lo Muzio, L.; Board, S.O.; Campisi, G. The Dental Management of Patients at Risk of Medication-Related Osteonecrosis of the Jaw: New Paradigm of Primary Prevention. *BioMed Res. Int.* **2018**, 1–10, doi:10.1155/2018/2684924.
2. Fusco, V.; Santini, D.; Armento, G.; Tonini, G.; Campisi, G. Osteonecrosis of jaw beyond antiresorptive (bone-targeted) agents: New horizons in oncology. *Expert Opin. Drug Saf.* **2016**, *15*, 925–935, doi:10.1080/14740338.2016.1177021.

3. Ruggiero, S.L.; Dodson, T.B.; Fantasia, J.; Goodday, R.; Aghaloo, T.; Mehrotra, B.; O'Ryan, F. American Association of Oral and Maxillofacial Surgeons, American Association of Oral and Maxillofacial Surgeons position paper on medication-related osteonecrosis of the jaw—2014 update. *J. Oral Maxillofac. Surg.* **2014**, *72*, 1938–1956, doi:10.1016/j.joms.2014.04.031.
4. Fedele, S.; Bedogni, G.; Scoletta, M.; Favia, G.; Colella, G.; Agrillo, A.; Bettini, G.; Di Fede, O.; Oteri, G.; Fusco, V.; et al. Up to a quarter of patients with osteonecrosis of the jaw associated with antiresorptive agents remain undiagnosed. *Br. J. Oral Maxillofac. Surg.* **2015**, *53*, 13–17, doi:10.1016/j.bjoms.2014.09.001.
5. Migliario, M.; Mergoni, G.; Vescovi, P.; Martino, I.; Alessio, M.; Benzi, L.; Renò, F.; Fusco, V. Osteonecrosis of the Jaw (ONJ) in Osteoporosis Patients: Report of Delayed Diagnosis of a Multisite Case and Commentary about Risks Coming from a Restricted ONJ Definition. *Dent. J.* **2017**, *5*, 13, doi:10.3390/dj5010013.
6. Bedogni, A.; Fedele, S.; Bedogni, G.; Scoletta, M.; Favia, G.; Colella, G.; Agrillo, A.; Bettini, G.; Di Fede, O.; Oteri, G.; et al. Staging of osteonecrosis of the jaw requires computed tomography for accurate definition of the extent of bony disease. *Br. J. Oral Maxillofac. Surg.* **2014**, *52*, 603–608, doi:10.1016/j.bjoms.2014.04.009.

© 2019 by the authors. Licensee MDPI, Basel, Switzerland. This article is an open access article distributed under the terms and conditions of the Creative Commons Attribution (CC BY) license (http://creativecommons.org/licenses/by/4.0/).

Extended Abstract
Ceaseless Oral Ulcerative Manifestations [†]

Vito Carlo Alberto Caponio [1], Carlo Lajolo [2], Giuseppe Troiano [1], Claudia Arena [1], Lorenzo Lo Muzio [1] and Michele Giuliani [1,*]

[1] Department of Clinical and Experimental Medicine, University of Foggia, 71122 Foggia, Italy; vito_caponio.541096@unifg.it (V.C.A.C.); giuseppe.troiano@unifg.it (G.T.); dadarena@hotmail.it (C.A.); lorenzo.lomuzio@unifg.it (L.L.M.)

[2] Institute of Dentistry and Maxillo-Facial Surgery, IRCCS Fondazione Policlinico "A. Gemelli", Università Cattolica del Sacro Cuore, 00168 Rome, Italy; carlo.lajolo@unicatt.it

* Correspondence: michele.giuliani@unifg.it; Tel.: +39-347-798-5261

[†] Presented at the XV National and III International Congress of the Italian Society of Oral Pathology and Medicine (SIPMO), Bari, Italy, 17–19 October 2019.

Published: 10 December 2019

On April 2016, a 24-year-old male patient came to our facilities for multiple mouth ulcers involving the tongue borders and the cheek mucosa (Figure 1). Patient didn't refer any systemic disease, but several episodes of diarrhea for a month; he was also diagnosed of incontinent cardia associated with hypersecretory antral gastropathy, HP negative. Blood test was slightly positive for anti-BP180 (20 U/mL—n.v. < 9) and negative for coeliac antibodies. Patient did not want to undergo biopsy because of the summer period but, because of the symptoms, he requested a therapy. Clobetasol propionate ointment and 3 mg Betamethasone tablets once a day were prescribed for a month with improvement in symptoms. While tapering, symptoms re-occurred. Few months later, after therapy suspension, patient underwent a triple biopsy of the lesions. New blood test was negative for pemphigoid and an altered CD4/CD8 ratio (0.9 vs. 1.4–1.9 n.v.) became evident. Report from biopsy was an epitheliotropic CD4+ lymphoproliferative disorder without clear monoclonality; a diagnosis of suspected lymphomatoid papulosis, without signs of cutaneous lesions and spontaneous remission, was hypothesized. [1]. The fecal calprotectin was 10 times higher (49,898 mg/Kg; n.v. 50 mg/Kg); although the biopsy of intestinal villi was negative, genetic typing of HLA locus DQ2/DQ8 [2], showed a predisposition for coeliac disease. Later, patient developed erythematous plaques involving the forearms and trunk, worsening after sun exposure. Abdomen MRI didn't show any alteration. Blood testing for HIV, Treponema pallidum, EBV, ASCA, ANCA, ANA and ENA were all negative. A year later, patient reported asymmetrical arthralgia, first in elbows, wrists and shoulders with muscular pain in flexor muscles of the forearms. Ventricular extrasystoles have been diagnosed recently: Holter ECG reported 30051 monomorphic isolated ventricular extrasystoles. The patient was visited by numerous specialist physicians, e.g. Dermatologist, Cardiologist, Rheumatologist but he is still undiagnosed (Figure 2) and finds improvement for oral ulcers with topical corticosteroids and occasionally systemic cortison (1 mg). Seloken (beta blocker) and Dibase 50.000 UI are also administered.

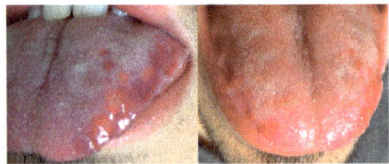

Figure 1. multiple mouth ulcers involving the tongue.

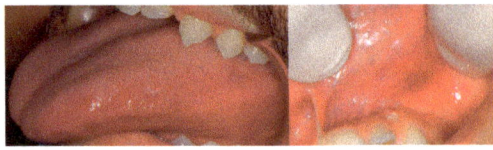

Figure 2. multiple mouth ulcers. Reticular lichenoid lesion (base and border of the tongue—4 years later).

Conflicts of Interest: The authors declare no conflict of interest.

References

1. Martinez-Cabriales, S.A.; Walsh, S. Lymphomatoid papulosis: an update and review. *J. Eur. Acad. Dermatol. Venereol.* **2019**, doi:10.1111/jdv.15931.
2. Lewis, D.; Haridy, J. Testing for coeliac disease. *Aust. Prescr.* **2017**, *40*, 105–108, doi:10.18773/austprescr.2017.029.

 © 2019 by the authors. Licensee MDPI, Basel, Switzerland. This article is an open access article distributed under the terms and conditions of the Creative Commons Attribution (CC BY) license (http://creativecommons.org/licenses/by/4.0/).

Extended Abstract

An Unusual Case of Oro-Facial Chronic Pain [†]

Dorina Lauritano [1],*, Alberta Lucchese [2] and Massimo Petruzzi [3]

[1] Department of Medicine and Surgery, Centre of Neuroscience of Milan, University of Milano-Bicocca, 20126 Milan, Italy
[2] Multidisciplinary Department of Medical-Surgical and Dental Specialties, University of Campania—Luigi Vanvitelli, 80138 Naples, Italy; alberta.lucchese@unicampania.it
[3] Interdisciplinary Department of Medicine, University of Bari, 70121 Bari, Italy; massimo.petruzzi@uniba.it
* Correspondence: dorina.lauritano@unimib.it; Tel.: +39-3356790163
[†] Presented at the XV National and III International Congress of the Italian Society of Oral Pathology and Medicine (SIPMO), Bari, Italy, 17–19 October 2019.

Published: 10 December 2019

1. Introduction

Fibromyalgia syndrome is a common form of diffuse musculoskeletal pain and fatigue (asthenia), which affects approximately 2% of world population.

The term fibromyalgia means pain in the muscles and fibrous connective structures (ligaments and tendons). This condition is called a "syndrome" because there are clinical signs and symptoms that are simultaneously present [1,2].

Fibromyalgia mainly affects the muscles and their insertions on the bones [3,4]. Although it may resemble an articular pathology, it is not arthritis and does not cause deformity of joint structures. Fibromyalgia is actually a form of extra-articular rheumatism or soft tissue.

The fibromyalgia syndrome lacks laboratory alterations. In fact, the diagnosis depends mainly on the symptoms the patient reports. Some people may consider these symptoms to be imaginary or unimportant. Over the past 10 years, however, fibromyalgia has been better defined through studies that have established guidelines for diagnosis. These studies have shown that certain symptoms, such as diffuse musculoskeletal pain, and the presence of specific algogenic areas for acupressure (tender points) are present in patients with fibromyalgia syndrome and not commonly in healthy people or in patients suffering from other painful rheumatic diseases. Tender point may be present at the level of temporo-mandibular junction and may overlap with temporo and cranio-mandibular disorders [5–8].

2. Therapeutically Approach

The anti-inflammatory drugs used to treat many rheumatic diseases do not show important effects in fibromyalgia. However, low doses of aspirin, ibuprofen and paracetamol can give some pain relief. Central analgesic drugs can reduce the painful symptoms of the fibromyalgic patient. Cortisones are ineffective and should be avoided due to their potential side effects. Drugs that facilitate deep sleep and muscle relaxation help many fibromyalgia patients to rest better. These drugs include tricyclic antidepressants (amitriptyline) and selective serotonin reuptake inhibitors (SSRIs) (paroxetine, fluoxetine) and other predominantly muscle relaxant drugs but structurally similar to antidepressants (cyclobenzaprine).

3. Exercise and Physical Therapies

Two of the main goals of fibromyalgia treatment are muscle stretching and training techniques for painful muscles and the gradual increase in cardiovascular (aerobic) fitness. Low or no impact aerobic activity, such as walking, cycling, swimming or exercising in water is generally the best way to start an exercise program.

4. Alternative Therapies

Even the so-called unconventional therapies such as dietary supplements or non-pharmacological treatments such as biofeed-back, acupuncture, gentle exercise and yoga can have positive effects on the symptoms of the fibromyalgic patient.

5. Case Report

39-year-old man suffered from significant chronic pain to arms, legs and extended to all body, non-restorative sleep, chronic fatigue, cutaneous rushes (Figure 1a,b). The chronic pain was referred to face, mouth and temporo-mandibular junction also. The patient was referred to a rheumatologist and then to a dermatologist. Blood testing was then performed: rheumatoid factor, anti-nuclear antibodies, anti-nDNA antibodies research, anti-ENA antibodies, and anti-cyclic citrullinated peptide antibodies were all negative. The dermatologist diagnosed atopic dermatitis. The patient was diagnosed with Fibromyalgia syndrome by the rheumatologist (FMS).

The onset of lower back pain, restless legs, and morning stiffness occurred one month before the visit. The symptoms were first described as severe pain. One-month later, lower back pain and legs pain forced the patient to bed rest. Similar symptoms of lower back pain during the following 2 months were attributed to the same cause. These episodes affected life quality and mobility, and forcing patient to bed.

The symptoms increased progressively and constantly. Migrant cutaneous rushes extended to hands, arms, neck, chest, legs and feet. A modest improvement was observed during summer and hot weather conditions, in particular during holydays. These had considerable impact on the everyday life, affecting social interaction and professional performance.

The ineffectiveness of pharmacological therapies in FMS came to patient's knowledge. The patient was prescribed selective serotonin–norepinephrine re-uptake inhibitor (SNRI) (i.e., paroxetine 10 mg die) and he referred a reduction in pain, cutaneous rushes and chronic fatigue.

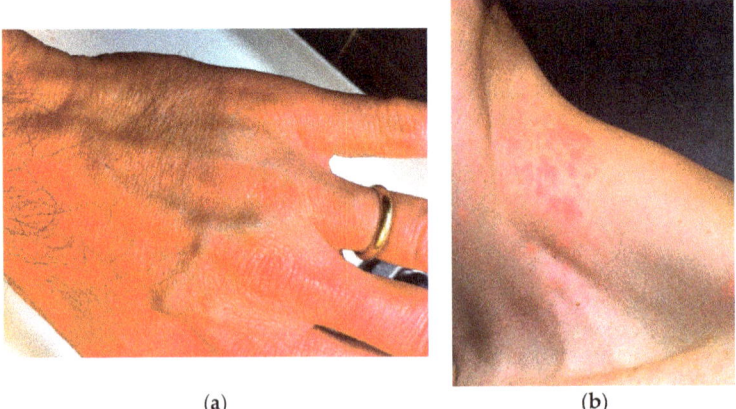

(a)   (b)

Figure 1. Rushes on skin of the (**a**) hand and (**b**) neck associated with fibromyalgia.

Conflicts of Interest: The authors declare no conflict of interest.

References

1. Balasubramaniam, R.; Laudenbach, J.M.; Stoopler, E.T. Fibromyalgia: An update for oral health care providers. *Oral Surg. Oral Med. Oral Pathol. Oral Radiol. Endodontol.* **2007**, *104*, 589–602.
2. Wolfe, F.; Smythe, H.A.; Yunus, M.B.; Bennett, R.M.; Bombardier, C.; Goldenberg, D.L.; Tugwell, P.; Campbell, S.M.; Abeles, M.; Clark, P.; et al. The American College of Rheumatology 1990 Criteria for the Classification of Fibromyalgia. *Arthritis Rheum. Off. J. Am. Coll. Rheumatol.* **1990**, *33*, 160–172.
3. Goulet, J.P.; Clark, G.T.; Flack, V.F.; Liu, C. The reproducibility of muscle and joint tenderness detection methods and maximum mandibular movement measurement for the temporomandibular system. *J. Orofac. Pain* **1998**, *12*, 17–26.
4. Leblebici, B.; Pektaş, Z.O.; Ortancil, O.; Hürcan, E.C.; Bagis, S.; Akman, M.N. Coexistence of fibromyalgia, temporomandibular disorder, and masticatory myofascial pain syndromes. *Rheumatol. Int.* **2007**, *27*, 541–544.
5. Plesh, O.; Wolfe, F.; Lane, N. The relationship between fibromyalgia and temporomandibular disorders: Prevalence and symptom severity. *J. Rheumatol.* **1996**, *23*, 1948–1952.
6. Korszun, A.; Papadopoulos, E.; Demitrack, M.; Engleberg, C.; Crofford, L. The relationship between temporomandibular disorders and stress-associated syndromes. *Oral Surg. Oral Med. Oral Pathol. Oral Radiol. Endodontol.* **1998**, *86*, 416–420.
7. Dao, T.T.; Reynolds, W.J.; Tenenbaum, H.C. Comorbidity between myofascial pain of the masticatory muscles and fibromyalgia. *J. Orofac. Pain* **1997**, *11*, 232–241.
8. Aaron, L.A.; Burke, M.M.; Buchwald, D. Overlapping conditions among patients with chronic fatigue syndrome, fibromyalgia, and temporomandibular disorder. *Arch. Intern. Med.* **2000**, *160*, 221–227.

© 2019 by the authors. Licensee MDPI, Basel, Switzerland. This article is an open access article distributed under the terms and conditions of the Creative Commons Attribution (CC BY) license (http://creativecommons.org/licenses/by/4.0/).

Extended Abstract

Mucocutaneous Leishmaniasis Mimicking Squamous Cell Carcinoma †

Rodolfo Mauceri [1,*], Giuseppina Campisi [1,2], Claudia Colomba [3], Olga Di Fede [1] and Vera Panzarella [2]

1. Department of Surgical, Oncological, and Oral Sciences, University of Palermo, 90 127 Palermo, Italy; campisi@odonto.unipa.it (G.C.); odifede@odonto.unipa.it (O.D.F.)
2. Oral Medicine and Dentistry for patients with special needs, AUOP "P. Giaccone" of Palermo, 90127 Palermo, Italy; panzarella@odonto.unipa.it
3. Department of Sciences for Health Promotion and Mother-Child Care, University of Palermo, 90127 Palermo, Italy; Claudia.colomba@unipa.it
* Correspondence: rodolfo.mauceri@unipa.it; Tel.: +39-09123864244
† Presented at the XV National and III International Congress of the Italian Society of Oral Pathology and Medicine (SIPMO), Bari, Italy, 17–19 October 2019.

Published: 10 December 2019

Leishmaniasis is a parasitic infectious disease caused by protozoan species belonging to the genus *Leishmania*. The leishmaniases are the third most important group of vectorborne diseases; it is endemic in 88 countries and it represents a major public health problem worldwide. In the Mediterranean area, most cases usually affect immunocompromised patients. Depending on the location of the lesions, the WHO described 11 human different clinical forms; the more common clinical forms can be categorized into four main groups: visceral leishmaniasis (VL); cutaneous leishmaniasis (CL); mucosal leishmaniasis (ML) and mucocutaneous leishmaniasis (MCL). Primary oral mucosal lesion can be the first sign of the infectious disease in all the clinical forms; usually oral lesions may be related to several symptoms (e.g., swallowing difficulties, dyspnea, dysphonia) [1–3]. In this case report, we presented a unusual case of MCL in a immunocompromised patients in South of Italy; whose presentation mimicked an oral squamous cell carcinoma (OSCC).

A Caucasian man was complaining about a cheeck lesion that began one-month prior the first visit appointment and was slowly involving the right lower lip. Patient reported to be affected by cutaneous psoriasis, in treatment with methotrexate and cyclosporine; additionally, he was a heavy smoker (Pack years: 26). Physical examination revealed a whitish, nodular and ulcerated lesion affecting the right oral commissure measuring approximately 2.5 × 1.5 cm, involving the lower lip right corner and buccal mucosa with hard consistency (Figure 1). Oral hygiene was poor, lower incisors were missing and mechanical trauma was present on the lesion. Palpation of the lesion revealed a firm consistency, the neck lymph nodes were slightly swelled.

Because a high suspicion of malignancy; the patient was included into a multidisciplinary care paths, named "GOTeC" ("Gruppo Oncologico Testa e Collo"—Policlinico "P. Giaccone" of Palermo), which deals with the entire diagnostic-therapeutic process of the patient suffering from H&N cancers, from the suspected diagnosis to therapeutic complete plans. Patient underwent a magnetic resonance imaging that highlighted an oval lesion measuring 1.2 × 2.9 × 2.1 cm (Figure 2), The neoformation extends anteriorly to the outer profile of the lower lip involving the orbicular and contiguous buccinator muscles, there was the evidence of swollen submandibular lymph nodes. Subsequently, an incisional biopsy was carried out, which showed non-specific chronic inflammation without neoplastic signs. At the post-surgical follow-up visit, the patient reported the evolution of the cutaneous psoriatic-like cutaneous lesions on the right harm, knees and especially foot towards a suppurative aspect. The patient was then referred to the dermatologic and infectious disease unit for further investigations; he underwent abdominal ultrasound imaging, serological test and a polymerase chain reaction test on skin mucosa biopsy samples to investigate the main causes of skin and soft-tissue Infections. Both results showed the leishmania infection, so the diagnosis of mucocutaneous leishmaniasis was established. Patient was pharmacologically treatment with liposomal amphotericin B and topical therapy for psoriasis; after discontinuation of immunosuppressive therapy. At the latest follow-up visit at 8 months, a complete remission of the oral and skin lesions was observed.

As well as for OSCC, it is important to emphasizes the importance of a multidisciplinary approach in the diagnosis and treatment of mucocutaneous leishmaniasis, whose diagnosis remains a challenge due to the heterogeneous clinical presentation. The diagnostic difficulties shown in this case were related to the development of such lesion in a patient who lived in a non-endemic area and the high suspicious presentation of the oral lesion for OSCC. Additionally, even if the Leishmaniasis usually affects immunocompromised patients, such as transplant recipients or those who are affected by HIV, it should be evaluated also in patients with long-term immunosuppressive pharmacological treatments.

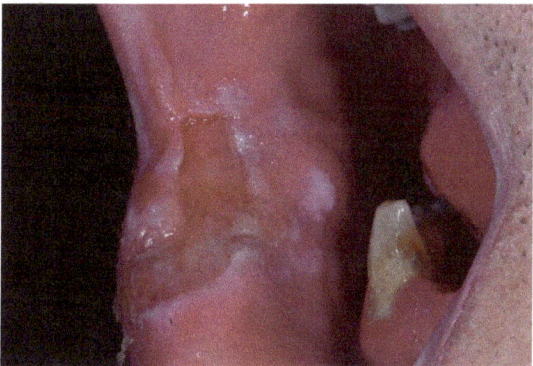

Figure 1. Clinical view of the oral lesion affecting the right labial commissure.

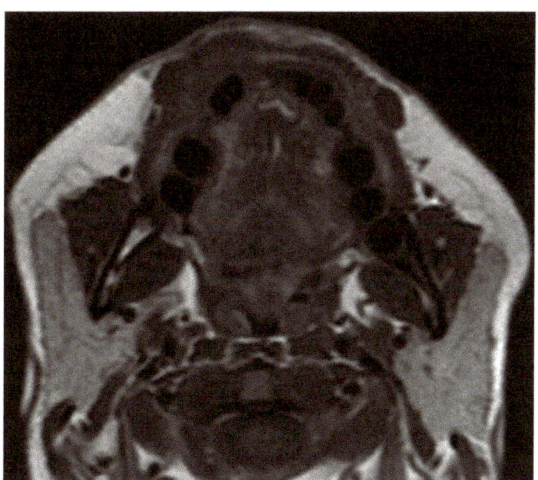

Figure 2. Magnetic resonance axial section.

Acknowledgments: This research received no external funding.

Conflicts of Interest: The authors declare no conflict of interest.

References

1. Mignogna, M.D.; Celentano, A.; Leuci, S.; Cascone, M.; Adamo, D.; Ruoppo, E.; Favia, G. Mucosal leishmaniasis with primary oral involvement: A case series and a review of the literature. *Oral Dis.* **2015**, *21*, e70–e78.
2. Mohammadpour, I.; Motazedian, M.H.; Handjani, F.; Hatam, G.R. Lip leishmaniasis: A case series with molecular identification and literature review. *BMC Infect. Dis.* **2017**, *17*, 96.
3. Ramos, A.; Múñez, E.; García-Domínguez, J.; Martinez-Ruiz, R.; Chicharro, C.; Baños, I.; Suarez-Massa, D.; Cuervas-Mons, V. Mucosal leishmaniasis mimicking squamous cell carcinoma in a liver transplant recipient. *Transpl. Infect. Dis.* **2015**, *17*, 488–492.

© 2019 by the authors. Licensee MDPI, Basel, Switzerland. This article is an open access article distributed under the terms and conditions of the Creative Commons Attribution (CC BY) license (http://creativecommons.org/licenses/by/4.0/).

Extended Abstract

Unusual Salivary Gland Tumor of the Palate: Clinical, Histological and Immunohistochemical Features [†]

Massimo Petruzzi [1,*], Fedora Della Vella [1], Pasquale Sportelli [2] and Eugenio Maiorano [3]

1. Interdisciplinary Department of Medicine (DIM), University of Bari "Aldo Moro", 70124 Bari, Italy; dellavellaf@gmail.com
2. Dental Clinic of "Azienda Universitario-Ospedaliera Policlinico di Bari", 70124 Bari, Italy; dr.pasqualesportelli@gmail.com
3. Department of Emergency and Organ Transplantation (DETO), University of Bari "Aldo Moro", 70124 Bari, Italy; eugenio.maiorano@uniba.it
* Correspondence: massimo.petruzzi@uniba.it; Tel.: +39-080-547-8388
† Presented at the XV National and III International Congress of the Italian Society of Oral Pathology and Medicine (SIPMO), Bari, Italy, 17–19 October 2019.

Published: 10 December 2019

1. Introduction

The latest WHO classification of salivary glands tumors includes more than 30 different benign and malignant histo-types [1]. Molecular genetic findings and immunohistochemistry have been integrated into the tumors profiles in order to obtain homogenous and reproducible diagnostic criteria. Nevertheless, the extreme clinical and morphological variability, does not always lead to conclusions that fully fit into the WHO parameters [2,3].

2. Case Presentation

A 36 years old man presented to our attention for a firm mass of the right hard palate (Figure 1a). The patient reported his dentist had noted the lesion about 12 months before. No pain or bleeding was reported. The clinical evaluation confirmed a tense-elastic, sessile nodule of 1.5 cm of diameter, covered by normal mucosa. Endodontic and periodontal evaluation, supported by radiograms (panoramic and intra-oral x-rays) excluded an abscessual origin. CT scan revealed a well-defined, circumscribed and captant mass, which caused bone saucerization but not complete palatal bone resorption. We decided to perform a diagnostic excisional biopsy. The neoplasm appeared macroscopically encapsulated and about 1 cm of diameter. The surgical wound healed with no complications. Microscopically, the capsule resulted partially incomplete with a focal intacapsular tumor localization. The myoepithelial cells were scarce and with plasmacytoid appearance. Neither marked cellular polymorphism nor atypical mitoses were noted on H&E. Immunohistochemical examination revealed positivity for CK7 and CK19 (this data excluded an adenoido-cystic adenoma) while CK14 was only weakly positive. PAS-diastase-stain was negative, thus ruling out acinic cell carcinoma. GFAP, usually overexpressed in polymorphous adenocarcinoma (PAC) and pleomorphic adenoma (PA) was only focally detectable. Myoepithelial markers, such as p40, p63, smooth muscle actin, myosin and calponin were not immunohistochemically evident. Absence of perineural invasion was confirmed by S100 protein immunostain stain. The percentage of Ki 67+ proliferating neoplastic cells was 3–4% (Figure 1b).

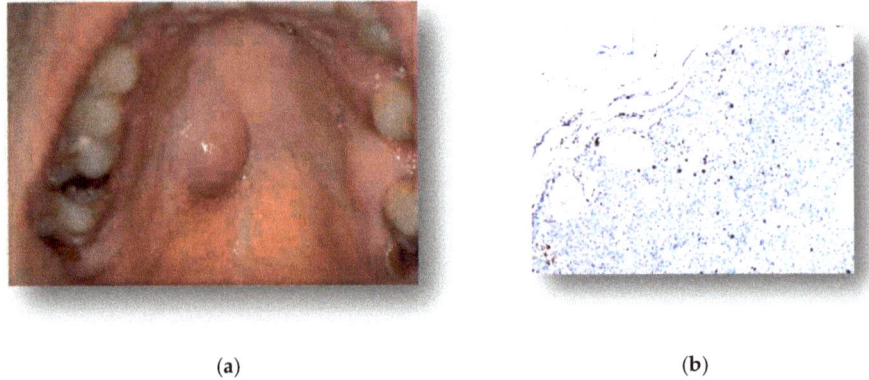

(a) (b)

Figure 1. (a) Clinical picture of the lesion; (b) k67 immunohistochemistry stain (10×).

3. Discussion

The current case recapitulates many of the problems that may arise in the differential diagnosis among the very different histotypes of salivary glands tumors. Partial capsular involvement by the tumor suggested a malignant neoplasm while the lack of atypical mitoses, perineural invasion, cellular atypicalities and necrosis supported for a benign lesion. The plausible differential diagnosis that we formulated included PA (with scarce myoepithelial component), PAC (ex "PLGA") and carcinoma ex-PA at an early stage. Due to the rarity of this kind of morpho-phenotypic features, further evaluation to possibly identify PLAG1 and HMGA2 genes mutations would be helpful to confirm a diagnosis in this still unsolved case.

Conflicts of Interest: The authors declare no conflict of interest.

References

1. IARC. *WHO Classification of Tumors*, 4th ed.; 2017; IARC press, Lyone, France, pp. 159–194.
2. Carlson, E.R.; Schlieve, T. Salivary Gland Malignancies. *Oral Maxillofac. Surg. Clin. N. Am.* **2019**, *31*, 125–144. doi:10.1016/j.coms.2018.08.007.
3. Galdirs, T.M.; Kappler, M.; Reich, W.; Eckert, A.W. Current aspects of salivary gland tumors—A systematic review of the literature. *GMS Interdiscip. Plast. Reconstr. Surg. DGPW* **2019**, *8*, doi:10.3205/iprs000138.

© 2019 by the authors. Licensee MDPI, Basel, Switzerland. This article is an open access article distributed under the terms and conditions of the Creative Commons Attribution (CC BY) license (http://creativecommons.org/licenses/by/4.0/).

Extended Abstract

Criticisms and Controversies in the Diagnosis of Cheilitis [†]

Umberto Romeo *, Federica Rocchetti and Alessandra Montori

Department of Oral and Maxillo-Facial Sciences, Sapienza University of Rome, 00161 Rome, Italy; federica.rocchetti@uniroma1.it (F.R.); montorialessandra@gmail.com (A.M.)
* Correspondence: umberto.romeo@uniroma1.it; Tel.: +393334134697
† Presented at the XV National and III International Congress of the Italian Society of Oral Pathology and Medicine (SIPMO), Bari, Italy, 17–19 October 2019.

Published: 11 December 2019

The term "cheilitis" refers to an inflammatory condition of the lip and encompasses an heterogeneous group of lesions of different etiology, clinical aspects, management and prognosis [1].

The lips cosmetically characterize the face of a person; they are fundamental to speach, eat and acting as a tattile organ. For such reasons, even a little alterations of the anatomy or functionality, are experienced by the patient with a certain discomfort and embarrassment.

Althought cheilitis has been identified and recognized for a long time, until now there have been confusion among clinicians due to poor knowledge, use of improper terminology, no clear recommendations for its work-up and classification [2].

Literature reports many papers on cheilitis but they are mostly case reports and overviews of therapeutic or diagnostic procedures based on personal experiences and results without specific criteria for classification.

In fact, about 30 articles were identified, performing a research using MEDLINE/PubMed and Cochrane Library databases with the keywords "cheilitis, lip cheilitis, lip inflammation, classification" and the Boolean operators "AND" and "OR" in order to combine the keywords,

Apart from that, some types of cheilitis requires a multidisciplinary approach between oral pathologists, dermatologists, otorhinolaryngologists, which additionally complicates adoption of a classification system.

According to our opinion, the classification proposed by Lugović-Mihić et al. based on the duration and etiology of cheilitis is useful, but should be integrate with some forms of cheilitis in order to guarantee simplification of the diagnosis, appropriate treatment and benefits for patients.

The classification is shown in Table 1.

Table 1. Lugović-Mihić et al. modified classification of cheilitis.

Mostly Reversible	Mostly Persistent	In Association with Dermatosis and Systemic Deseases (Common Deseases)
<ul><li>Cheilitis simplex</li><li>Angular/Infective cheilitis</li><li>Contact/Eczematous cheilitis</li><li>Exfoliative cheilitis</li><li>Drug related cheilitis</li><li>Post–radio and/or chemotherapy cheilitis</li></ul>	<ul><li>Actinic cheilitis</li><li>Granulomatous cheilitis</li><li>Glandular cheilitis</li><li>Plasma cell cheilitis</li></ul>	<ul><li>Lupus erythematosus</li><li>Lichen Planus</li><li>Angioedema</li><li>Pemphigus/Pemphigoid</li><li>Xerostomia</li><li>Erythema multiforme</li><li>Crohn's desease</li><li>Sarcoidosis</li><li>Psoriasis, etc.</li></ul>

Mostly reversible and irreversible cheilitis are competence of stomatologists; instead, for the cheilits connected to dermatoses and systemic diseases the multidisciplinary care is mandatory.

Cheilitis of more clinical interest will be treated below.

Cheilitis simplex is one of the most common types and presents as cracked, fissured, desquamated lips. Frequent lip licking or sucking promote dryness and irritation, ending in separation of the mucosa and cracking. Therapy mostly involves advice on dealing with environmental conditions, application of lip balms and sometimes topical corticosteroids.

Angular cheilitis, also called *perlèche*, typically manifests at the corners of the mouth. Vitamin and mineral deficiencies (B vitamins, iron, zinc, etc.), the loss of skin turgor due to aging, the loss of vertical dimension of the face due to severe edentulous states, retrognathic malocclusion are the most frequent causes [3]. Concurrence of bacterial (staphylococcal and betahemolytic streptococcal) or candidal infectionsis common. Therapy includes elimination of predisposing factors and often topical antimycotics, antiseptics and sometimes corticosteroids.

Contact cheilitis (Figure 1) is caused by the irritating or allergic effects of various substances found in many products such as lipsticks, oral hygiene products (toothpastes), food (e.g., eggs and crustaceans), spices or flavouring agents and dental materials [4]. Patch tests can play a pivotal role in the diagnosis, although its diagnostic use is still controversial in the literature [5].

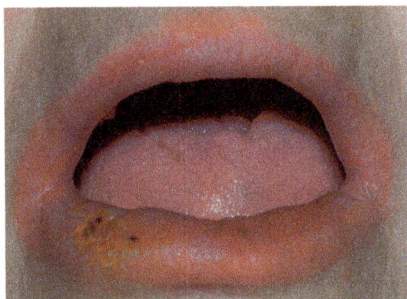

Figure 1. Contact cheilitis.

Post –radio and/or –chemotherapy cheilitis is a lips localization of oral mucositis that may occur in 30–60% of patients receveing radiotherapy for head-neck cancer, 90% of patients receiving concomitant chemoterapy and in 75% of patients receveing high-dose of myeloablative drugs used for conditioning regimens for allogenic hematopoietic stemcell transplantation. The lesion undergo spontaneous resolution.

Actinic cheilitis (Figure 2) is due to prolonged sun exposure and occurs almost exclusively in fair-skinned people who work outdoors or spend too much time in the sun. It occurs more frequently on the lower lip, probably because the site is more directly exposed to sunlight. Clinical aspects consist in painless whitish discolorations at the border of the lips and the skin. The lips may gradually become scaly and indurated and chronic focal ulcerations can occur. Actinic cheilitis is a potential malignant disorder: the malignant transformation ranges from 10% to 30% [4]. Smoking might produce a synergistic effect. Follow-up is recommended with the execution of incisional biopsies in lesions with more severe clinical aspects. Surgery treatment options include cryosurgery, electrosurgery, laser vaporization and scalpel vermilionectomy. In a systematic review conducted by Vasconcelos Carvalho et al. remission rate and recurrence rate of actinic cheilitis were higher and lower respectively for surgical treatment compared to non surgical treatment [6].

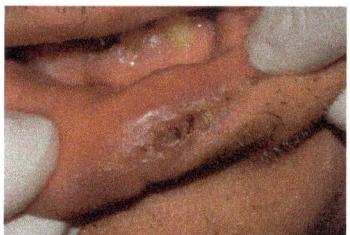

Figure 2. Actinic cheilitis.

Cheilitis granulomatosa (of Miescher) is a rare, painless and idiophatic swelling of the lip belonging to the larger group of orofacial granulomatosis caused by non-caseating granulomatous inflammation. When accompanied by facial palsy and plicated tongue, it is referred as Melkersson–Rosenthal syndrome. The association between cheilitis granulomatosa and Crohn's disease has been documented widely in the literature: sometimes cheilitis granulomatosa may precede Crohn's disease by up to several years. Therapy may include topical, intralesional and systemic corticosteroids/antibiotics and surgery in severely disfiguring cases [7].

Plasmacellular cheilitis is a rare entities and manifests tipically on the lower lip with a circumscribed erosive and erithematous plaques or patches accompanied by pricking. The etiology is unknow and histologically is characterized by dense plasmacells infiltration in dermis. The clinical examination frequently leads to an initial misdiagnosis, which requires biopsy. Treatment include intralesional or systemic steroid and local application of immunomodulatory agents (e.g., tacrolimus, pimecrolimus) [8].

With the term *cheilitis associated with systemic diseases* refers to the lip manifestations of diseases such lupus erythematosus, oral lichen planus, pemphigus, pemphigoid, erytema multiforme, etc. We would like to suggest the introduction in this category of peri-oral psoriasis. Clinicians must pay attentions because sometimes lip lesions are the first manifestation of psoriasis, without skin involvement [9].

Angioedema results from vascular leakage due to release of vasoactive mediators such as histamine, serotonin, and bradykinin with extravasations of fluid into the superficial tissues causing edema. It can be classified as allergic, pseudo-allergic, or non-allergic atopic eczema.

Knowledge gaps amongst clinicians about different forms of cheilitis may influence the misdiagnosis and the mismanagement. Complete medical history, physical examination, appropriate diagnostic workup and interdisciplinary collaboration are the key factors in recognizing the right type of cheilitis and of its successful treatment. A small number of lip lesions, however, are potentially lethal and thus require appropriate intervention to prevent further morbidity and mortality.

Funding: No funding has been received in support of this paper.

Conflicts of Interest: The authors declare no conflict of interest.

References

1. Collet, E.; Jeudy, G.; Dalac, S. Cheilitis, perioraldermatitis and contactallergy. *Eur. J. Dermatol.* **2013**, *23*, 303–307.
2. Lugović-Mihić, L.; Pilipović, K.; Crnarić, I.; Šitum, M.; Duvančić, T. Differential Diagnosis of Cheilitis— How to Classify Cheilitis? *Acta Clin. Croat.* **2018**, *57*, 342–351
3. Bakula, A.; Lugović-Mihić, L.; Šitum, M.; Turčin, J.; Šinković, A. Contact allergy in the mouth: diversity of clinical presentations and diagnosis of common allergens relevant to dental practice. *Acta Clin. Croat.* **2011**, *50*, 553–561.
4. Mortazavi, H.; Baharvand, M.; Mehdipour, M. Oral Potentially Malignant Disorders: An Overview of More than 20 Entities. *J. Dent. Res. Dent. Clin. Dent. Prospect.* **2014**, *8*, 6.

5. Minciullo, P.L.; Paolino, G.; Vacca, M.; Gangemi, S.; Nettis, E. Unmet diagnostic needs in contact oral mucosal allergies. *Clin. Mol. Allergy* **2016**, *14*, 10.
6. Carvalho, M.V.; de Moraes, S.L.D.; Lemos, C.A.A.; Santiago Júnior, J.F.; Vasconcelos, B.C.D.E.; Pellizzer, E.P. Surgical versus non-surgical treatment of actinic cheilitis: A systematic review and meta-analysis. *Oral. Dis.* **2019**, *25*, 972–981.
7. Critchlow, W.A.; Chang, D. Cheilitis granulomatosa: A review. *Head Neck Pathol.* **2014**, *8*, 209–213.
8. Lee, J.Y.; Kim, K.H.; Hahm, J.E.; Ha, J.W.; Kwon, W.J.; Kim, C.W.; Kim, S.S. Plasma Cell Cheilitis: A Clinicopathological and Immunohistochemical Study of 13 Cases. *Ann. Dermatol.* **2017**, *29*, 536–542.
9. Martí, N.; Pinazo, I.; Revert, A.; Jordá, E. Psoriasis of the lips. *J. Dermatol. Case Rep.* **2009**, *3*, 50–52.

© 2019 by the authors. Licensee MDPI, Basel, Switzerland. This article is an open access article distributed under the terms and conditions of the Creative Commons Attribution (CC BY) license (http://creativecommons.org/licenses/by/4.0/).

Extended Abstract

Oral Leukoplakia. A More Challenging Disorder than It Seems †

Serban Radu Tovaru

Oral Medicine Discipline, Faculty of Dental Medicine, "Carol Davila" University of Medicine and Pharmacy, 050037 Bucharest, Romania; serban.tovaru@gmail.com; Tel.: +40-072-303-0739

† Presented at the XV National and III International Congress of the Italian Society of Oral Pathology and Medicine (SIPMO), Bari, Italy, 17–19 October 2019.

Published: 11 December 2019

Treatment of oral leukoplakia remains largely a debatable problem [1]. According to widely accepted opinions the diagnosis of oral leukoplakia should have two steps [2]: the provisory diagnosis should include general patient evaluation (sex, age) and a local thorough oral exam: clinical form, unilocular, multilocular, topographical site. Other diseases that could produce similar lesions should be eliminated following a specific protocol. In the same time, any local favoring factors (smoking, alcohol, other habits, trauma, infections, etc.) should be eliminated. After 3–4 months a definitive diagnosis will be put accordingly [2]. In some instances the lesions could considerably diminish or even disappear (Figures 1 and 2). If the lesions persist a histopathological evaluation with degree of dysplasia, immunological markers and genetic tests are recommended. The degree of dysplasia should be evaluated according to the WHO scale or according to the binary system [3,4]. If possible genetic tests of the surrounding area of the lesion are recommended. Brush cytology is the best method to be used. Following that complex evaluation protocol, the risk of malignisation could be higher or lower.

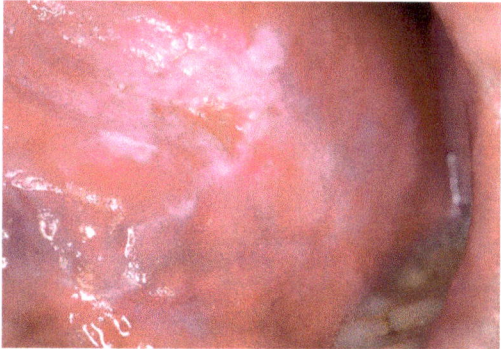

Figure 1. 2015 Male, 70 years, smoker, drinker, Candida positive. **Clinical Diagnosis**: Non-homogenous diffuse ulcerated lingual leukoplakia. **Treatment:** Quit smoking, Antifungal treatment. Biopsy.

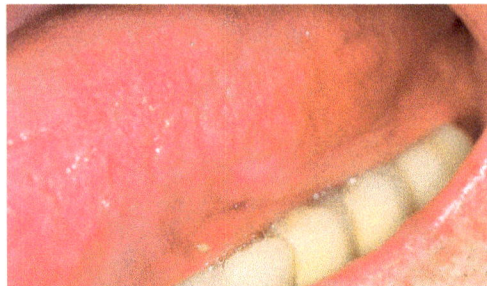

Figure 2. Same patient. 2019. Clinical aspect. The patient quit smoking; 6 months later residual keratotic lesions were removed. No sign of relapse

Lesions with high degree of dysplasia, genetic disturbances (Loss of heterozigosy), multilocular type, speckled, location in high-risk zones should be eliminated. Surgical methods should be adapted according to the width, length, and position of the lesion keeping in mind the borders of oncological security. Classic scalpel removal, laser surgery, cryotherapy should be adapted in accordance with the clinical and histopathological data. A significant percentage of relapse is to be expected in case of idiopathic leukoplakia [1]. A post surgical, strict clinical follow-up must be instituted.

In the same time, some of the patients do not accept a surgical treatment or simply don't come back for control and parts of the lesions evolve without treatment. In this group some of the lesions disappear spontaneously, other enlarge, other shrink and other turn to cancer [5]. Thus, the wait-and-see attitude is an option and a reality. Patients included in this group should be closely followed and if the clinical aspect changes, new biopsies are recommended and a surgical treatment becomes an option. In our opinion that is a valid medical option.

No unified therapeutical guidelines exist so far [1].

Conflicts of Interest: The authors declare no conflict of interest.

References

1. Holmstrup, P.; Dabelsteen, E. Oral leukoplakia—To treat or not to treat. *Oral Dis.* **2016**, *22*, 494–497.
2. Van der Waal, I. Potentially malignant disorders of the oral and oropharyngeal mucosa; terminology, classification and present concepts of management. *Oral Oncol.* **2009**, *45*, 317–323.
3. Speight, P.M.; Khurram, S.A.; Kujan, O. Oral potentially malignant disorders: Risk of progression to malignancy. *Oral Surg. Oral Med. Oral Pathol. Oral Radiol.* **2018**, *125*, 612–627.
4. Brouns, E.R.A.; Baart, J.A.; Bloemena, E.; Karagozoglu, H.; Van der Waal, I. The relevance of uniform reporting in oral leukoplakia: Definition, certainty factor and staging based on experience with 275 patients. *Med. Oral Patol. Oral y Cir. Bucal* **2013**, *18*, e19.
5. Kuribayashi, Y.; Tsushima, F.; Morita, K.I.; Matsumoto, K.; Sakurai, J.; Uesugi, A.; Sato, K.; Oda, S.; Sakamoto, K.; Harada, H. Long-term outcome of non-surgical treatment in patients with oral leukoplakia. *Oral Oncol.* **2015**, *51*, 1020–1025.

© 2019 by the authors. Licensee MDPI, Basel, Switzerland. This article is an open access article distributed under the terms and conditions of the Creative Commons Attribution (CC BY) license (http://creativecommons.org/licenses/by/4.0/).

Extended Abstract

Persistent Oral Erosion: A Diagnostic Challange †

Matteo Val *, Melania Lupatelli, Marco Ardore, Roberto Marino and Monica Pentenero

Department of Oncology, Unit of Oral Medicine and Oral Oncology, University of Turin, San Luigi Gonzaga Hospital, 001171 Orbassano, Italy; melanialupatelli@gmail.com (M.L.); ardore.marco@gmail.com (M.A.); roberto.marino@unito.it (R.M.); monica.pentenero@unito.it (M.P.)
* Correspondence: matteo.val@outlook.it; Tel.: +39-3402313851
† Presented at the XV National and III International Congress of the Italian Society of Oral Pathology and Medicine (SIPMO), Bari, Italy, 17–19 October 2019.

Published: 11 December 2019

The diagnosis and treatment of oral erosion/ulceration is often challenging due to the clinician's limited exposure to the conditions that may cause the lesions and their similar appearances. While many oral ulcers and erosions are the result of chronic trauma, some may indicate an underlying systemic condition such as a gastrointestinal dysfunction, malignancy, immunologic abnormality, or cutaneous disease. Correctly establishing a definitive diagnosis is of major importance to clinicians who manage patients with oral mucosal disease [1].

A 77-year-old male edentulous former smoker (stopped in 1962) suffering of hypertension, angina and glaucoma in pharmacological treatment, was admitted to our Oral Medicine and Oral Oncology Clinic on January 2010 with severe and persistent eroded lesions noticed 3 months prior to his visit (Figure 1A). Lesions involved the left buccal mucosa; extending to left labial commissure and the vermillion. No other lesions could be detected at an oral and cutaneous examination. Map biopsies under Velscope® guide were performed but the pathological assessment didn't highlight any alteration in the oral mucosa. Miconazole 2% gel (3 applications per day per 15 days) and Chlortetracycline 3% ointment (3 applications per day per 15 days) treatment improved the clinical aspect of the lesion. After 3 months since the erosion didn't show up any more improvement, the patient underwent cryosurgery. A nearly complete regression of the erosion was obtained (Figure 1B). On March 2011, meanwhile two new denture were performed, a worsening of the clinical aspect and of the symptoms had led to the execution of new biopsy samples, which highlighted the presence of mild-moderate dysplasia (Figure 1C). The patient decided to do not perform the excision, so a new biopsy was performed on November 2011(Figure 1D). The pathological assessment showed the presence of an in situ carcinoma. The patient underwent photodynamic therapy in 2012. During the follow-up on March 2013 (Figure 1E) a new severe worsening appeared, with the involvement of the buccal mucosa and also of the skin of the labial commissure by erosion with scabs. The histologic examination ruled out a relapse. Until 2016 the erosion changed several time his clinical aspect. Hematological and Quantiferon exams were required, but they excluded a Tuberculosis infection and a lymphoproliferative disorder. Several pharmacological treatments (antimicotic and antibiotic topical use) couldn't heal the lesion, biopsies were repeated in the years (September 2013 (Figure 1F,G), March 2014) until April 2016 (Figure 1H,I) when the pathological assessment disclosed a mild dysplasia. Because of the worsening of the systemic condition of the patient (since 2015 the patient was suffering of Alzheimer disease), according with the son of the patient, was decided to keep only clinical follow-up. On August 2017 the patient died due to a complication of senile dementia.

Diagnosis of oral erosion-ulceration can be challenging and requires careful clinical examination and history taking. Persistent oral ulcerations can result from a variety of disparate etiologies and, therefore, may pose a diagnostic challenge. Historically, the diagnosis of many of these conditions has been made based only on clinical presentation, sometimes along with tissue biopsy [2]. However, not every case will present the typical clinical or histological features associated with a certain condition. The management of many of these conditions has been similarly nonspecific, often involving non-targeted anti-inflammatory or immunosuppressive agents.

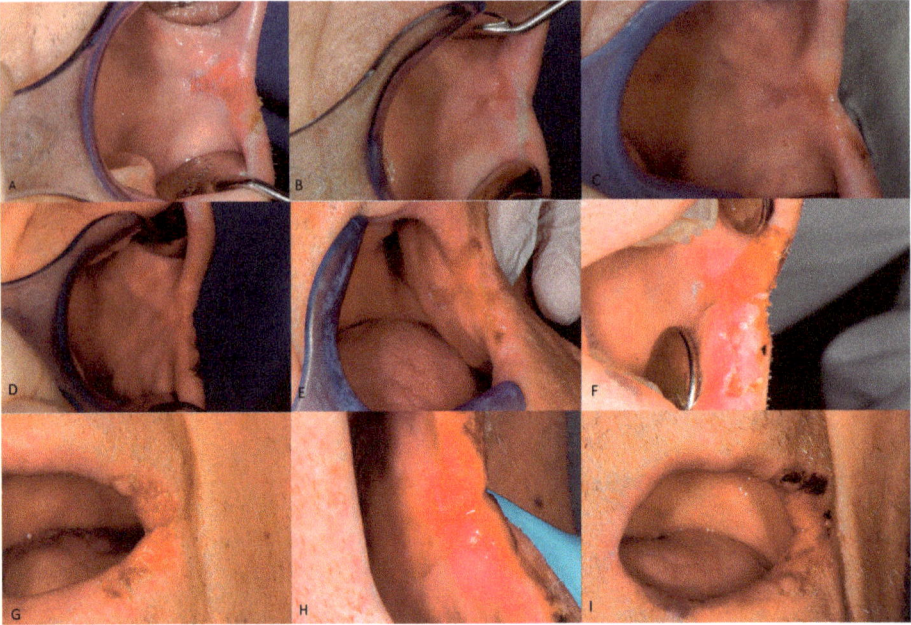

Figure 1. (**A**) Clinical aspects of the buccal mucosa lesion at first examination January 2010. (**B**) Erosion healing after cryotherapy on June 2010. (**C**) Worsening of the erosion on May 2011 with a pathological assessment of mild-moderate dysplasia. (**D**) November 2011 in situ carcinoma. (**E**) After photodinamic therapy a relapse of the erosion was highlighted on March 2013 with a pathological assessment of dense inflammatory infiltrate of plasma cells without dysplasia. (**F,G**) September 2013: worsening of the lesion with a skin's involvement, biopsies ruled out the presence of dysplasia. (**H,I**) April 2016: Enlargement of the lesion with the presence of mild dysplasia.

Conflicts of Interest: The authors declare no conflict of interest.

References

1. Bilodeau, E.A.; Lalla, R.V. Recurrent oral ulceration: Etiology, classification, management, and diagnostic algorithm. *Periodontol 2000* **2019**, *80*, 49–60.
2. Siu, A.; Landon, K.; Ramos, D.M. Differential diagnosis and management of oral ulcers. *Semin. Cutan. Med. Surg.* **2015**, *34*, 171–177.

 © 2019 by the authors. Licensee MDPI, Basel, Switzerland. This article is an open access article distributed under the terms and conditions of the Creative Commons Attribution (CC BY) license (http://creativecommons.org/licenses/by/4.0/).

Extended Abstract

A Case of Intra-Oral Bone Exposure of the Hard Palate: A Clinical Diagnostic Dilemma †

Davide B Gissi *, Andrea Gabusi and Lucio Montebugnoli

Department of Biomedical and Neuromotor Sciences, Section of Oral Sciences, University of Bologna, 40125 Bologna, Italy; andrea.gabusi3@unibo.it (A.G.); lucio.montebugnoli@unibo.it (L.M.)
* Correspondence: davide.gissi@unibo.it; Tel.: +39-0512088123
† Presented at the XV National and III International Congress of the Italian Society of Oral Pathology and Medicine (SIPMO), Bari, Italy, 17–19 October 2019.

Published: 11 December 2019

Osteonecrosis or aseptic and avascular necrosis of the jaws (ONJ), has been defined as an area of intra-oral and/or extra-oral exposed bone and may result in significant patient morbidity. In the last period osteonecrosis of the jaws related to Bisphosphonate (BRONJ) and, recently, Denosumab related osteonecrosis of the jaws (DRONJ) and other medicaments such as antiangiogenic agents (MRONJ) have been abundantly reported [1], but different etiologies may cause a maxillo-mandibular osteonecrosis. In the present report we describe the case of a patient affected by Crohn's Disease referred to Department of Biomedical and Neuromotor Sciences, unit of Oral Medicine, University of Bologna on September 2108 for the presence of a painful lesion in the hard palate. His medical history revealed a diagnosis of Crohn's disease in 2016. Medical treatment with mesalazine, prednisone and later with adalimumab and infliximab didn't significantly improve the clinical condition of the patient. In June 2018 a surgical resection of intestinal tract affected by Crohn's disease was performed. Finally, in July 2018 a laboratory test revealed a positivity for HLA B51, a genetic marker related to Behcet's disease.

In the clinical examination we revealed the presence of a 3–4 cm of painful bone exposure in the right side of hard palate and presence of a generalized severe periodontitis (Figure 1a,b). In this patient various local or systemic phenomena might take part in osteonecrosis of the oral cavity. ONJ may be related to adalimumab and/or infliximab therapy, 2 anti TNF-α antibody used to treat Crohn's disease and other autoimmune diseases. Few authors reported a relationship between infliximab and/or adalimumab therapy and appearance of ONJ [2–4]. In this particular case the area of exposed bone didn't appear after a surgical odontostomatological procedure, but the patient showed a severe generalized periodontitis, a well known ONJ risk factor. Chronic osteomyelitis, a rare infection of the medullary portion of the jawbone with purulent exudate related with immunocompromised conditions of the patient, was the second considered hypothesis. Finally, oral manifestation of systemic condition was the third hypothesis, even if an intra-oral osteonecrosis is not a typical oral manifestation Crohn related.

Different clinical and instrumental investigations were performed. An incisional biopsy in the hard palate was scheduled. At histology, Hematoxylin Eosin stained sections revealed acanthosis, hyperkeratosis and the presence of a dense trans-mural non necrotizing granulomatous chronic inflammation. PAS staining methods and additional colorations with Ziehl Neelsen and Giemsa did not reveal fungal, bacterial or parasite presence. A CT scan showed a massive maxillary bone loss in the right hard palate region without maxillary sinus involvement. Clinical investigations were also performed with the aim to reconsider the initial diagnosis of Crohn's disease. Indeed, HLA-B51 positivity is strongly related to Behcet's Disease [5] and when the gastrointestinal tract is involved, a differential diagnosis between Behcet's Disease and Crohn's Disease may be very difficult [6]. Clinical systemic investigations revealed multiple ulcers in ipo-pharingeal and laryngeal tract and a painful purulent ulceration on the skin, diagnosed as Pyoderma gangrenosum. At the same time, the patient didn't show ocular and genital ulcers and pathergy test was negative. Unexpectedly, in the 18th of October 2018 patient presented a fatal cardiovascular complication and died. Parents refused the autopsy.

Unfortunately, in this case the formulation of a definitive diagnosis consistent with the clinical presentation and a consequent treatment plan was not possible. Diagnosis of multisystemic inflammatory conditions is based on clinical findings and different Crohn's disease manifestations are similar to Behcet's Disease manifestations. Diagnosis of MRONJ is also based on clinical findings and few informations still exist about the exact relationship between biological agents and ONJ. In the future presence of specific pathognomonic laboratory tests may be fundamental in cases with multiple comobordities to avoid a delay in diagnosis and treatment.

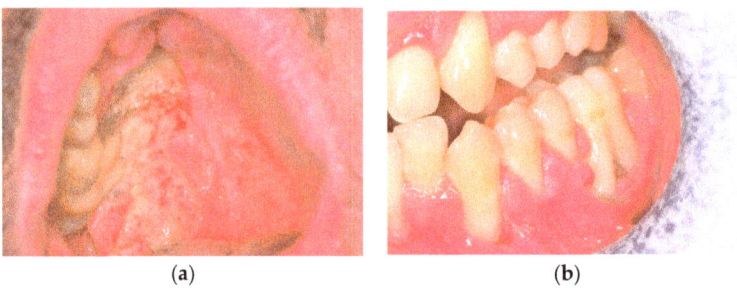

(a) (b)

Figure 1. Presence of intra-oral bone exposure in the right side of the hard palate (**a**) and presence of a severe generalized periodontitis (**b**).

References

1. King, R.; Tanna, N.; Patel, V. Medication-related osteonecrosis of the jaw unrelated to bisphosphonates and denosumab-a review. *Oral. Surg. Oral Med. Oral. Pathol. Oral. Radiol.* **2019**, *127*, 289–299.
2. Favia, G.; Tempesta, A.; Limongelli, L.; Crincoli, V.; Iannone, F.; Lapadula, G.; Maiorano, E. A Case of Osteonecrosis of the Jaw in a Patient with Crohn's Disease Treated with Infliximab. *Am. J. Case Rep.* **2017**, *18*, 1351–1356.
3. Cassoni, A.; Romeo, U.; Terenzi, V.; Della Monaca, M.; Rajabtork Zadeh, O.; Raponi, I.; Fadda, M.T.; Polimeni, A.; Valentini, V. Adalimumab: Another Medication Related to Osteonecrosis of the Jaws? *Case Rep. Dent.* **2016**, *2016*, 2856926.
4. Preidl, R.H.M.; Ebker, T.; Raithel, M.; Wehrhan, F.; Neukam, F.W.; Stockmann, P. Osteonecrosis of the jaw in a Crohn's disease patient following a course of Bisphosphonate and Adalimumab therapy: A case report. *BMC Gastroenterol* **2014**, *14*, 6.
5. Demirseren, D.D.; Ceylan, G.G.; Akoglu, G.; Emre, S.; Erten, S.; Arman, A.; Metin, A. HLA-B51 subtypes in Turkish patients with Behçet's disease and their correlation with clinical manifestations. *Genet. Mol. Res.* **2014**, *13*, 4788–4796.
6. Valenti, S.; Gallizzi, R.; De Vivo, D.; Romano, C. Intestinal Behçet and Crohn's disease: two sides of the same coin. *Pediatr. Rheumatol. Online J.* **2017**, *15*, 33.

© 2019 by the authors. Licensee MDPI, Basel, Switzerland. This article is an open access article distributed under the terms and conditions of the Creative Commons Attribution (CC BY) license (http://creativecommons.org/licenses/by/4.0/).

Extended Abstract

Evaluation of *Echinophora Tenuifolia* L. Extracts on HSC-2 Cell Line [†]

Claudia Arena [1,*], Marco Vairano [1], Marco Mascitti [2], Andrea Santarelli [2], Mario Dioguardi [1] and Khrystyna Zhurakivska [1]

1. Department of Clinical and Experimental Medicine, University of Foggia, 71122 Foggia, Italy; marcovairano@gmail.com (M.V.); mario.dioguardi@unifg.it (M.D.); khrystyna.zhurakivska@unifg.it (K.Z.)
2. Department of Clinical Specialistic and Dental Sciences, Marche Polytechnic, 60131 Ancona, Italy; marcomascitti86@hotmail.it (M.M.); andrea.santarelli@univpm.it (A.S.)
* Correspondence: claudia.arena@unifg.it; Tel.: +39-3406305228
† Presented at the XV National and III International Congress of the Italian Society of Oral Pathology and Medicine (SIPMO), Bari, Italy, 17–19 October 2019.

Published: 11 December 2019

Oral squamous cell carcinoma survival is still poor, although the improvement in treatments both in surgical and chemotherapy [1]. Conventional treatments are also associated with acute and chronic toxicity, which leads to a decrease of quality life [2]. Natural compounds are a promising alternative, since they can affect the different steps of tumor cell development, with poor toxic events. *E. Tenuifolia* L. showed promising anticancer activities in several cancer models, due to its antioxidant activity [3,4]. In order to verify its anticancer activity in a cellular model of squamous cell carcinoma, HSC-2, we performed a MTT cell proliferation assay, by adding this compound at time 0 (t0). HSC-2 were cultured in cell culture flask at 37 °C with 5% CO_2 in RPMI medium supplemented with 10% fetal bovine serum (FBS), 1% penicillin-streptomycin and 1% L-Glutammine. We used the inflorescences total extract (20 mg/mL in EtOH 70%) and we evaluated the production of formazan in order to determine the cell viability, at 24 h, 48 h and 72 h. When comparing these results, with results coming from cells in addition with ethanol only, we observed a decrease in cell viability of 40%. Future studies should deeply investigate the role of *Echinophora Tenuifolia* L. as possible adjuvant in cancer therapy, in order to understand its role in pathways and molecular targets.

Conflicts of Interest: The authors declare no conflict of interest.

References

1. Suh, Y.; Amelio, I. Clinical update on cancer: molecular oncology of head and neck cancer. *Cell Death Dis.* **2014**, *5*, e1018.
2. Valdez, J.A.; Brennan, M.T. Impact of Oral Cancer on Quality of Life. *Dent. Clin. N. Am.* **2018**, *62*, 143–154.
3. Marrelli, M.; Pisani, F.; Amodeo, V.; Duez, P.; Conforti, F. *Echinophora tenuifolia* L. branches phytochemical profile and antiproliferative activity on human cancer cell lines. *Nat. Prod. Res.* **2019**, 1–4, doi:10.1080/14786419.2018.1548457
4. Marrelli, M.; Statti, G.A.; Menichini, F.; Conforti, F. *Echinophora tenuifolia* L. inflorescences: phytochemistry and in vitro antioxidant and anti-inflammatory properties in LPS-stimulated RAW 264.7 macrophages. *Plant Biosyst.* **2017**, *151*, 1073–1081.

© 2019 by the authors. Licensee MDPI, Basel, Switzerland. This article is an open access article distributed under the terms and conditions of the Creative Commons Attribution (CC BY) license (http://creativecommons.org/licenses/by/4.0/).

Extended Abstract

Preventing Bacterial Leakage in Implant-Abutment Connection: A Review [†]

Dorina Lauritano [1],*, Giulia Moreo [1], Francesco Carinci [2], Alberta Lucchese [3], Dario di Stasio [3], Fedora della Vella [4] and Massimo Petruzzi [4]

1. Department of Medicine and Surgery, Centre of Neuroscience, University of Milano-Bicocca, 20126 Milan, Italy; moreo.giulia@gmail.com
2. Department of Morphology, Surgery and Experimental Medicine, University of Ferrara, 44121 Ferrara, Italy; crc@unife.it
3. Multidisciplinary Department of Medical-Surgical and Dental Specialties, University of Campania—Luigi Vanvitelli, 80138 Naples, Italy; alberta.lucchese@unicampania.it (A.L.); dario.distasio@unicampania.it (D.d.S.)
4. Interdisciplinary Department of Medicine, University of Bari, 70121 Bari, Italy; fdellavella@gmail.com (F.d.V.); massimo.petruzzi@uniba.it (M.P.)
* Correspondence: dorina.lauritano@unimib.it; Tel.: +39-335-679-0163
† Presented at the XV National and III International Congress of the Italian Society of Oral Pathology and Medicine (SIPMO), Bari, Italy, 17–19 October 2019.

Published: 10 December 2019

1. Introduction

Osseointegration can be affected by oral conditions, in particular the micro gap at the implant-abutment-connection (IAC) represents a site for dental plaque aggregation favoring bacterial leakage that can increase inflammatory cells at the level of the IAC, causing peri-implantitis [1]. This micro gap, once early colonized, may constitute a bacterial reservoir, that could subsequently contaminate fixture's surroundings and interfere with peri-implant tissues health [2,3].

2. Aim

The aim of this review is to describe, according to the most recent literature, the different kind of implant-abutment connection and their ability to reduce bacterial leakage and thus preventing peri-implantitis.

3. Materials and Methods

The following database were consulted: Pubmed (n = 26), Scopus (n = 90), Research gate (n = 7) and were found a total of 123 articles. Duplicates were excluded and after reading abstract and titles, were excluded those articles that were off topic. The remaining ones (n = 24) were assessed for full-text elegibility: we excluded 5 articles because they were case report, 2 because there was no clear reference to the relationship IAC and bacterial leakage and 2 because was not pertinent with the argument. Fifteen articles were included in the review.

4. Results and Conclusions

From the review, it's clear that exists a relationship between the IAC and bacterial leakage. All the connection presented an amount of micro-gap and bacterial micro-leakage but conical and mixed connection systems seem to behave better. Moreover, both connections seem to have a better load's distribution and mixed one has also anti-rotational properties very useful during the positioning of the prosthesis [4].

Conflicts of Interest: The authors declare no conflict of interest.

References

1. Albrektsson, T.; Zarb, G.; Worthington, P.; Eriksson, A.R. The long-term efficacy of currently used dental implants: A review and proposed criteria of success. *Int. J. Oral Maxillofac. Implants* **1986**, *1*, 11–25.
2. Ottria, L.; Lauritano, D.; Andreasi Bassi, M.; Palmieri, A.; Candotto, V.; Tagliabue, A.; Tettamanti, L. Mechanical, chemical and biological aspects of titanium and titanium alloys in implant dentistry. *J. Biol. Regul. Homeost. Agents* **2018**, *32*, 81–90
3. De Oliveira, D.P.; Ottria, L.; Gargari, M.; Candotto, V.; Silvestre, F.J.; Lauritano, D. Surface modification of titanium alloys for biomedical application: From macro to nano scale. *J. Biol. Regul. Homeost. Agents* **2017**, *31* (Suppl. 1), 221–232.
4. Canullo, L.; Penarrocha-Oltra, D.; Soldini, C.; Mazzocco, F.; Penarrocha, M.; Covani, U. Microbiological assessment of the implant-abutment interface in different connections: cross-sectional study after 5 years of functional loading. *Clin. Oral Implant Res.* **2015**, *26*, 426–434

© 2019 by the authors. Licensee MDPI, Basel, Switzerland. This article is an open access article distributed under the terms and conditions of the Creative Commons Attribution (CC BY) license (http://creativecommons.org/licenses/by/4.0/).

Extended Abstract

Effects of Periodontal Therapy on the Management of Cardiovascular Disease [†]

Dorina Lauritano [1],*, Giulia Moreo [1], Francesco Carinci [2], Alberta Lucchese [3], Dario di Stasio [3], Fedora della Vella [4] and Massimo Petruzzi [4]

1. Department of Medicine and Surgery, University of Milano-Bicocca, Centre of neuroscience, 20126 Milan, Italy; moreo.giulia@gmail.com
2. Department of Morphology, Surgery and Experimental medicine. University of Ferrara, 44121 Ferrara, Italy; crc@unife.it
3. Multidisciplinary Department of Medical-Surgical and Dental Specialties, University of Campania—Luigi Vanvitelli, 80138 Naples, Italy; alberta.lucchese@unicampania.it (A.L.); dario.distasio@unicampania.it (D.d.S.)
4. Interdisciplinary Department of Medicine, University of Bari, 70121 Bari, Italy; fdellavella@gmail.com (F.d.V.); massimo.petruzzi@uniba.it (M.P.)
* Correspondence: dorina.lauritano@unimib.it; Tel.: +39-3356790163
† Presented at the XV National and III International Congress of the Italian Society of Oral Pathology and Medicine (SIPMO), Bari, Italy, 17–19 October 2019.

Published: 10 December 2019

Cardiovascular disease (CVD) is a common cause of death, representing 29% of the mortality all over the word. Estimates for 2006 show that CVD is one of the world's main cause of death, with 17.1 million death per year. More than 70 million Americans have been diagnosed with various forms of CVD, including high blood pressure, coronary artery disease (acute myocardial infarction and angina pectoris), disorders of peripheral arteries etc. There is strong evidence that periodontal disease (PD) is associated with an increased risk of CVD. In addiction many patients with CVD are also affected by PD, which can be mild or severe [1,2]. The aim of this manuscript is to investigate the effects of periodontal therapy on the management of CVD. 34 randomised controlled trials and reviews were included in this mauscript to test the effects of different periodontal therapies for patients with CVD. In conclusion we may affirm that there is some lack of knowledge on relations between PD and CVD, however there is sufficient evidence to justify a periodontal treatment to prevent CVD, in fact PD is very prevalent in middle-aged population and can have a significant impact on the cardiovascular function [3,4].

Conflicts of Interest: The authors declare no conflict of interest.

References

1. American Academy of Periodontology. International workshop for classification of periodontal diseases and conditions. *Ann. Periodontol.* **1999**, *4*, 7–112.
2. Blum, A.; Front, E.; Peleg, A. Periodontal care may improve systemic inflammation. *Clin. Investig. Med.* **2007**, *30*, E114–E117.
3. Socransky, S.S.; Haffajee, A.D.; Cugini, M.A.; Smith, C.; Kent, R.L., Jr. Microbial complexes in subgingival plaque. *J. Clin. Periodontol.* **1998**, *25*, 134–144.
4. Luis, A.J. Atherosclerosis. *Nature* **2000**, *407*, 233–241.

© 2019 by the authors. Licensee MDPI, Basel, Switzerland. This article is an open access article distributed under the terms and conditions of the Creative Commons Attribution (CC BY) license (http://creativecommons.org/licenses/by/4.0/).

Extended Abstract

Two-Way Relationship between Diabetes and Periodontal Disease: A Reality or a Paradigm? [†]

Dorina Lauritano [1],*, Giulia Moreo [1], Francesco Carinci [2], Alberta Lucchese [3], Dario di Stasio [3], Fedora della Vella [4] and Massimo Petruzzi [4]

1. Department of Medicine and Surgery, Centre of Neuroscience, University of Milano-Bicocca, 20126 Milan, Italy; moreo.giulia@gmail.com
2. Department of Morphology, Surgery and Experimental Medicine, University of Ferrara, 44121 Ferrara, Italy; crc@unife.it
3. Multidisciplinary Department of Medical-Surgical and Dental Specialties, University of Campania—Luigi Vanvitelli, 80138 Naples, Italy; alberta.lucchese@unicampania.it (A.L.); dario.distasio@unicampania.it (D.d.S.)
4. Interdisciplinary Department of Medicine, University of Bari, 70121 Bari, Italy; fdellavella@gmail.com (F.d.V.); massimo.petruzzi@uniba.it (M.P.)
* Correspondence: dorina.lauritano@unimib.it; Tel.: +39-335-679-0163
† Presented at the XV National and III International Congress of the Italian Society of Oral Pathology and Medicine (SIPMO), Bari, Italy, 17–19 October 2019.

Published: 10 December 2019

Diabetes mellitus (DM) and periodontal disease (PD) are both chronic diseases.

From one side, DM have an adverse effect on PD, and on the other side PD may influence DM. Systemic therapy of DM with glycaemic control, affects the progress of PD. Reversely treatment of PD combined with the administration of systemic antibiotics seems to have a double effect on diabetic patients reducing the periodontal infection and improving the glycaemic control [1,2].

Inflammation, altered host responses, altered tissue homeostasis are common characteristic of both DM and PD. The potential common pathophysiologic pathways of direct or reverse relationship of DM and PD are still unknown and further in vitro and in vivo studies are needed to explore this relationship [3,4].

Conflicts of Interest: The authors declare no conflict of interest.

References

1. Centers for Disease Control and Prevention. *National Diabetes Fact Sheet: National Estimates and General Information on Diabetes and Prediabetes in the United States, 2011*; Department of Health and Human Services, Centers for Disease Control and Prevention: Atlanta, GA, USA, 2011. Available online: http://www.cdc.gov/diabetes/pubs/pdf/ndfs_2011. pdf (accessed on 6 January 2013).
2. Chen, L.; Magliano, D.J.; Zimmet, P.Z. The worldwide epidemiology of type 2 diabetes mellitus–present and future perspectives. *Nat. Rev. Endocrinol.* **2012**, *8*, 228–236.
3. Albandar, J.M.; Rams, T.E. Global epidemiology of periodontal diseases: An overview. *Periodontol. 2000* **2002**, *29*, 7–10.
4. Monnier, V.M.; Mustata, G.T.; Biemel, K.L.; Reihl, O.; Lederer, M.O.; Zhenyu, D.A.I.; Sell, D.R. Cross-linking of the extracellular matrix by the Maillard reaction in aging and diabetes: An update on "a puzzle nearing resolution". *Ann. N. Y. Acad. Sci.* **2005**, *1043*, 533–544.

© 2019 by the authors. Licensee MDPI, Basel, Switzerland. This article is an open access article distributed under the terms and conditions of the Creative Commons Attribution (CC BY) license (http://creativecommons.org/licenses/by/4.0/).

Extended Abstract
Helicobacter Pylory and Oral Diseases [†]

Dorina Lauritano [1,*], Giulia Moreo [1], Francesco Carinci [2], Alberta Lucchese [3], Dario di Stasio [3], Fedora della Vella [4] and Massimo Petruzzi [4]

1. Department of Medicine and Surgery, Centre of Neuroscience, University of Milano-Bicocca, 20126 Milan, Italy; moreo.giulia@gmail.com
2. Department of Morphology, Surgery and Experimental Medicine, University of Ferrara, 44121 Ferrara, Italy; crc@unife.it
3. Multidisciplinary Department of Medical-Surgical and Dental Specialties, University of Campania—Luigi Vanvitelli, 80138 Naples, Italy; alberta.lucchese@unicampania.it (A.L.); dario.distasio@unicampania.it (D.d.S.)
4. Interdisciplinary Department of Medicine, University of Bari, 70121 Bari, Italy; fdellavella@gmail.com (F.d.V.); massimo.petruzzi@uniba.it (M.P.)
* Correspondence: dorina.lauritano@unimib.it; Tel.: +39-335-679-0163
† Presented at the XV National and III International Congress of the Italian Society of Oral Pathology and Medicine (SIPMO), Bari, Italy, 17–19 October 2019.

Published: 10 December 2019

Helicobacter pylori (*H. pylori*) gastric infection is considered one of the most common human infections. It occurs in half of the world's population is the most common cause of adenocarcinoma of the distal stomach [1].

The risk in developing gastric cancer is believed to be related to differences among *H. pylori* strains and the inflammatory responses mediated by host genetic factors.

The accepted evidence is that the *H. pylori* strains reach the stomach by ingestion through the mouth, and because of its non-invasive nature, the stomach is the definitive site for colonization [2,3].

One of the key issues related to the eradication of gastric *H. pylori* has been the importance of oral hygiene and periodontal procedures. Dental plaque control and periodontal therapy can prevent gastric *H. pylori* infection recurrence for patients with gastric diseases associated with *H. pylori* [4].

Conflicts of Interest: The Authors declare no conflict of interest.

References

1. Atherton, J.C.; Cao, P.; Peek, R.M., Jr.; Tummuru, M.K.; Blaser, M.J.; Cover, T.L. Mosaicism in vacuolating cytotoxin alleles of Helicobacter pylori. Association of specific vacA types with cytotoxin production and peptic ulceration. *J. Biol. Chem.* **1995**, *270*, 17771–17777.
2. Cave, D.R. Transmission and epidemiology of Helicobacter pylori. *Am. J. Med.* **1996**, *100*, 12S–17S.
3. Covacci, A.; Telford, J.L.; Del Giudice, G.; Parsonnet, J.; Rappuoli, R. Helicobacter pylori virulence and genetic geography. *Science* **1999**, *284*, 1328–1333.
4. Dunn, B.E.; Cohen, H.; Blaser, M.J. Helicobacter pylori. *Clin. Microbiol. Rev.* **1997**, *10*, 720–741.

© 2019 by the authors. Licensee MDPI, Basel, Switzerland. This article is an open access article distributed under the terms and conditions of the Creative Commons Attribution (CC BY) license (http://creativecommons.org/licenses/by/4.0/).

Extended Abstract

The Effect of Tobacco Smoking on Periodontal Health [†]

Dorina Lauritano [1],*, Giulia Moreo [1], Francesco Carinci [2], Alberta Lucchese [3], Dario di Stasio [3], Fedora della Vella [4] and Massimo Petruzzi [4]

1. Department of Medicine and Surgery, Centre of Neuroscience, University of Milano-Bicocca, 20126 Milan, Italy; moreo.giulia@gmail.com
2. Department of Morphology, Surgery and Experimental Medicine, University of Ferrara, 44121 Ferrara, Italy; crc@unife.it
3. Multidisciplinary Department of Medical-Surgical and Dental Specialties, University of Campania—Luigi Vanvitelli, 80138 Naples, Italy; alberta.lucchese@unicampania.it (A.L.); dario.distasio@unicampania.it (D.d.S)
4. Interdisciplinary Department of Medicine, University of Bari, 70121 Bari, Italy; fdellavella@gmail.com (F.d.V.); massimo.petruzzi@uniba.it (M.P.)
* Correspondence: dorina.lauritano@unimib.it; Tel.: +39-335-679-0163
† Presented at the XV National and III International Congress of the Italian Society of Oral Pathology and Medicine (SIPMO), Bari, Italy, 17–19 October 2019.

Published: 10 December 2019

Periodontal diseases (PD) affect about half of the adult population all over the world. PD is caused by bacterial infection inducing an inflammatory response with progressive destruction of the periodontal tissues and finally the lost of teeth. Tobacco smoking (TS), alcohol consumption, and systemic conditions such as diabetes, osteoporosis, malnutrition and stress are considered additional risk factors [1,2]. This short review examines the potential causal association between TS and PD. There are many studies for a higher level of PD among smokers. Greater levels of clinical alveolar bone loss, tooth mobility, probing pocket depth and tooth loss are more frequent in smokers than in non-smokers. The modification of the periodontitis micro-flora in smokers influences the development of PD. Also, there are data suggesting smoking effects on both host responses in humans. Response to periodontal treatment is different in smokers and non-smokers. Various clinical studies have demonstrated that TS is a major risk factor for poor response to periodontal therapy. TS is a factor that has the potential to negatively affect healing and the outcome of implant treatment [3]. It is mandatory for dentists and dental hygienist to promote smoking-cessation programs as well as educate our community on the benefits of not smoking [3,4].

Conflicts of Interest: The authors declare no conflict of interest.

References

1. American Academy of Periodontology. International workshop for classification of periodontal diseases and conditions. *Ann. Periodontol.* **1999**, *4*, 7–112.
2. Blum, A.; Front, E.; Peleg, A. Periodontal care may improve systemic inflammation. *Clin. Investig. Med.* **2007**, *30*, E114–E117.
3. Samet, J.M.; Wipfli, H.L. Globe still in grip of addiction. *Nature* **2010**, *463*, 1020–1021.
4. Jha, P.; Chaloupka, F.J.; Corrao, M.; Jacob, B. Reducing the burden of smoking world-wide: Effectiveness of interventions and their coverage. *Drug Alcohol Rev.* **2006**, *25*, 597–609.

© 2019 by the authors. Licensee MDPI, Basel, Switzerland. This article is an open access article distributed under the terms and conditions of the Creative Commons Attribution (CC BY) license (http://creativecommons.org/licenses/by/4.0/).

Extended Abstract

Predicting Death in Patients with Squamous Cell Carcinoma of the Tongue [†]

Vito Carlo Alberto Caponio [1,*], Giuseppe Troiano [1], Marco Mascitti [2], Andrea Santarelli [2], Rodolfo Mauceri [3] and Lorenzo Lo Muzio [1]

1. Department of Clinical and Experimental Medicine, University of Foggia, 71122 Foggia, Italy; giuseppe.troiano@unifg.it (G.T.); lorenzo.lomuzio@unifg.it (L.L.M.)
2. Department of Clinical Specialistic and Dental Sciences, Marche Polytechnic, 60131 Ancona, Italy; marcomascitti86@hotmail.it (M.M.); andrea.santarelli@staff.univpm.it (A.S.)
3. Department of Surgical, Oncological and Oral Sciences, University of Palermo, 90127 Palermo, Italy; rodolfo.mauceri@unipa.it
* Correspondence: vito_caponio.541096@unifg.it; Tel.: +39-380-742-1416
† Presented at the XV National and III International Congress of the Italian Society of Oral Pathology and Medicine (SIPMO), Bari, Italy, 17–19 October 2019.

Published: 10 December 2019

Tongue squamous cell carcinoma (TSCC) accounts for 40% of all squamous cell carcinoma involving the mucosal surface of the oral cavity. TSCC is highly invasive and aggressive and, nowadays, TNM staging system is considered the gold standard in predicting patients' outcomes. Nevertheless, patients with tumors classified under the same TNM stage, can undergo different outcomes, with differences in behavior, aggressiveness and therapy response. This panorama calls for new biomarkers which could be used in clinical practice in a precision medicine point of view [1]. TCGA Database [2] offers wide genome sequencing analysis, including mRNA expression of whole genome. We included patients with TSCC and we downloaded data of mRNA expression per patient and survival outcome [3]. We performed a differential expression analysis and ranking according to the outcome. This bioinformatics approach led to the detection of 12 promising genes which resulted to be able to predict patients' prognosis. Poor is known about the detected genes and future studies are needed to test this gene panel in order to assess the accuracy in predicting death in an external cohort of patients. In addition, since these genes could be linked to a higher risk of patient's death, they could be actor in some pathological pathways, which characterize the tumor biology, for example in chemotherapy resistance, metastasis or tumor growth [4].

Conflicts of Interest: The authors declare no conflict of interest.

References

1. Moeckelmann, N.; Ebrahimi, A. Prognostic implications of the 8th edition American Joint Committee on Cancer (AJCC) staging system in oral cavity squamous cell carcinoma. *Oral Oncol.* **2018**, *85*, 82–86.
2. TCGA Database. Available online: https://www.cancer.gov/tcga (accessed on 25 November 2019).
3. Grossman Robert, L.; Heath Allison, P. Toward a Shared Vision for Cancer Genomic Data. *N. Engl. J. Med.* **2016**, *375*, 1109–1112.
4. Almangush, A.; Heikkinen, I.; Heikkinen, I.; Makitie, A.A.; Coletta, R.D.; Läärä, E.; Leivo, I.; Salo, T. Reply to Comment on Prognostic biomarkers for oral tongue squamous cell carcinoma: A systematic review and meta-analysis. *Br. J. Cancer* **2017**, *117*, 856–866.

© 2019 by the authors. Licensee MDPI, Basel, Switzerland. This article is an open access article distributed under the terms and conditions of the Creative Commons Attribution (CC BY) license (http://creativecommons.org/licenses/by/4.0/).

Extended Abstract

Role of Periodontal Therapy Plus Sodium Doxycycline in the Management of Desquamative Gingivitis: A Pilot Study [†]

Mario Carbone *, Veronica Avolio, Marco Cabras, Paola Carcieri, Paolo Giacomo Arduino and Roberto Broccoletti

Department of Surgical Sciences, Oral Medicine Section, CIR-Dental School, University of Turin, 10126 Turin, Italy; veronicaavolio18@gmail.com (V.A.); cabrasmarco300@gmail.com (M.C.); carcieri.paola@libero.it (P.C.); paologiacomo.arduino@unito.it (P.G.A.); roberto.broccoletti@unito.it (R.B.)
* Correspondence: mario_carbone@libero.it; Tel.: +39-011-633-15-22
† Presented at the XV National and III International Congress of the Italian Society of Oral Pathology and Medicine (SIPMO), Bari, Italy, 17–19 October 2019.

Published: 11 December 2019

1. Aim

Desquamative gingivitis (DG) is a clinical representation of many autoimmune diseases, presenting with erythema, shedding and ulceration of free and attached gingiva. In the last decade, evidence suggested that DG could play a role in increasing the long-term risk for periodontal tissue breakdown, leading to the inclusion of periodontal therapies as additional treatment for DG patients [1]. Topical doxycycline gel has been tested successfully as additional option to scaling and root planing (SRP), triggering the reduction of periodontal pathogens [2]. Aim of the present work was to assess the role of 14% doxycycline gel (Ligosan®, Dental Trey, Turin, Italy) as an additional tool to non-surgical periodontal treatment on patients affected by Oral Lichen Planus (OLP)-related DG, and active periodontitis.

2. Methods

Patients referred to the Department of Oral Medicine, CIR Dental School, Turin, were enrolled. Inclusion criteria were: symptomatic DG, caused by histologically confirmed OLP, and active periodontitis, with at least three periodontal intraosseous defects of 5 mm or more. The eligible patients were randomly assigned either to "cases", where Ligosan® was administered with SRP, or to "controls", undergoing SRP alone. Severity of DG and periodontitis were evaluated through the following parameters: visual analogue scale (VAS), full-mouth plaque score (FMPS), full-mouth bleeding score (FMBS), and desquamative gingivitis clinical score (DGCS) [3]. Statistical analysis was performed through Wilcoxon and signed rank test. Table 1 shows the main characteristics of each therapeutic protocol.

Table 1. Treatment designed for Cases and Controls.

Timing	Cases	Controls
T0	Anamnesis; oral examination and orthopantomography (OPT); photographic documentation; fulfilment of periodontal chart; evaluation of VAS, FMBS, FMPS, DGCS	Anamnesis; oral examination and orthopantomography (OPT); photographic documentation; fulfilment of periodontal chart; evaluation of VAS, FMBS, FMPS, DGCS
T1 (one week after T0)	Ultrasonic scaling; instructions for domiciliary oral hygiene; Chlorexidine 0.20% mouthrinse three times daily for one month	Ultrasonic scaling; instructions for domiciliary oral hygiene; Chlorexidine 0.20% mouthrinse three times daily for one month
T2 (one week after T1)	Root planing of maxillary teeth; reinforcement of motivation for oral hygiene	Root planing of maxillary teeth; reinforcement of motivation for oral hygiene
T3 (one week after T2)	Root planing of mandibular teeth; application of Ligosan® in the maxillary intraosseous pockets; no flossing and sole occlusal brushing for 7 days; soft diet for 3 days	Root planing of mandibular teeth; reinforcement of motivation for oral hygiene
T4 (one week after T3)	Application of Ligosan® in the mandibular intraosseous pockets; no flossing and sole occlusal brushing for 7 days; soft diet for 3 days	Conventional oral and radiographic (OPT) examination; photographic documentation; fulfilment of periodontal chart; re-evaluation of VAS, FMBS, FMPS, DGCS
T5 (eight weeks after T4)	conventional oral and radiographic (OPT) examination; photographic documentation; fulfilment of periodontal chart; re-evaluation of VAS, FMBS, FMPS, DGCS	/

3. Results

Ten patients were recruited: six cases (4 M; 2 F; mean age: 54 years old; range 48–84), and four controls (1 M; 3 F; mean age: 72 years old; range 64–74). Among cases, statistical analysis revealed a significant decrease of FMPS and FMBS ($p < 0.05$), a quasi-significant decrease of DGCS ($p = 0.07$), with no significant variations of VAS ($p > 0.05$), after treatment. Controls experienced no significant fluctuations for either of the aforesaid parameters ($p > 0.05$). When compared, only FMPS was significantly decreased in cases rather than controls ($p < 0.05$).

4. Conclusions

To our knowledge, this is the first attempt of experimenting the reliability of topical doxycycline in patients with DG and active periodontitis. The main limitation of this work relies in the very small sample of patients recruited. Ligosan® showed encouraging results in the reduction of DG severity, and improvement of the periodontal status. Further studies on larger samples could be useful to establish the consistency of these preliminary results.

Acknowledgments: The Authors would like to thank Dental Trey for providing free Ligosan® samples.

Conflicts of Interest: The Authors declare no conflict of interest. Dental Trey had no role in the design of the study; in the collection, analyses, or interpretation of data; in the writing of the manuscript.

References

1. Guiglia, R.; Di Liberto, C.; Pizzo, G.; Picone, L.; Lo Muzio, L.; Gallo, P.D.; Campisi, G.; D'Angelo, M. A combined treatment regimen for desquamative gingivitis in patients with oral lichen planus. *J. Oral Pathol. Med.* **2007**, *36*, 110–116.

2. Ratka-Krüger, P.; Schacher, B.; Bürklin, T.; Böddinghaus, B.; Holle, R.; Renggli, H.H.; Eickholz, P.; Kim, T.S. Non-surgical periodontal therapy with adjunctive topical doxycycline: A double-masked, randomized, controlled multicenter study. II. Microbiological results. *J. Periodontol.* **2005**, *76*, 66–74.
3. Arduino, P.G.; Broccoletti, R.; Sciannameo, V.; Scully, C. A practical clinical recording system for cases of desquamative gingivitis. *Br. J. Dermatol.* **2017**, *177*, 299–301.

© 2019 by the authors. Licensee MDPI, Basel, Switzerland. This article is an open access article distributed under the terms and conditions of the Creative Commons Attribution (CC BY) license (http://creativecommons.org/licenses/by/4.0/).

Extended Abstract

Primordial Odontogenic Tumour: A Systematic Review [†]

Fabio Croveri [1,*], Vittorio Maurino [2], Alessandro d'Aiuto [1], Marta Dani [1], Andrea Boggio [1] and Lorenzo Azzi [1]

1. Department of Medicine and Surgery, University of Insubria, ASST Sette Laghi, Dental Clinic, Unit of Oral Medicine and Pathology, 21100 Varese, Italy; daiuto.alex@gmail.com (A.d.); marta.dani92.md@gmail.com (M.D.); andreaboggio90@gmail.com (A.B.); lorenzoazzi86@hotmail.com (L.A.)
2. Department of Medicine and Surgery, University of Insubria, ASST Sette Laghi, Dental Clinic, Unit of Pediatric Dentistry, 21100 Varese, Italy; vittorio.maurino@gmail.com
* Correspondence: Fabio.croveri@icloud.com; Tel.: +39-0332825623
† Presented at the XV National and III International Congress of the Italian Society of Oral Pathology and Medicine (SIPMO), Bari, Italy, 17–19 October 2019.

Published: 11 December 2019

1. Introduction

Primordial odontogenic tumor (POT) is a novel entity that was described in 2014 and that is included in the group of benign mixed epithelial and mesenchymal odontogenic tumors [1].

In recent years, several papers have added new cases with some clinical and histopathological aspects that slightly differ from those described in the original report, or that had not been previously observed.

The aim of this systematic review is to update all available data on POT published in the literature and to identify those features of the neoplasm that require further investigation.

2. Materials and Methods

A systematic review of literature was conducted using PubMed, Embase, Web of Science and Scopus and following PRISMA guidelines [2]. J-Stage, LILACS and Google Scholar were also checked. Publications reporting cases with enough clinicopathological information were included, without any time or language restrictions. Histopathological, immunohistochemical or radiological studies were considered for qualitative analysis.

3. Results

A total of 28 publications were included (Figure 1). Sixteen papers were used for quantitative analysis (Figure 2; Table 1). A total of 17 cases of POT were identified in the literature. POT is a benign mixed epithelial and mesenchymal odontogenic tumour composed of a loose fibrous connective tissue resembling the dental papilla. The distinguishing feature of the neoplasm is the presence of a single layer of columnar cells resembling the inner enamel epithelium of the developing tooth that entirely covers the lesion peripherally.

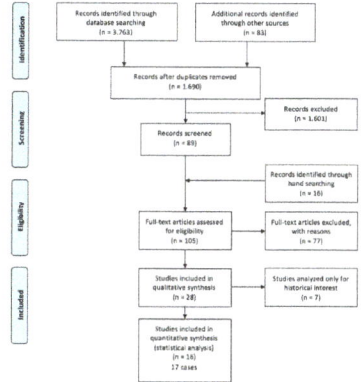

Figure 1. Study screening process.

Figure 2. Ages of patients with POT. There were three subpopulations: infants in the deciduous dentition stage, children in a mixed dentition stage and adolescent/young adults in the permanent dentition stage.

Table 1. Epidemiological, clinical, radiographic, macroscopic and therapeutic features of published POT cases.

† One of the six original cases was reported in the Basque Country, Spain, European Union; ‡ Event though tha paper was published in 2018, the cases were originaly described in two abstracts in 2016.

4. Conclusions

Some issues about POT remain unclear and require future reports.

The description of the odontogenic epithelium covering the ectomesenchyme was often contradictory, while it remains debatable whether peripheral ameloblastic islands or hard dental tissue deposition can occasionally occur within the tumor [3,4].

Conflicts of Interest: The authors declare no conflict of interest.

References

1. Mosqueda-Taylor, A.; Pires, F.R.; Aguirre-Urízar, J.M.; Carlos-Bregni, R.; de la Piedra-Garza, J.M.; Martínez-Conde, R.; Martínez-Mata, G.; Carreño-Álvarez, S.J.; da Silveira, H.M.; de Barros Dias, B.S.; et al. Primordial odontogenic tumour: Clinicopathological analysis of six cases of a previously undescribed entity. *Histopathology* **2014**, *65*, 606–612, doi:10.1111/his.12451.
2. Liberati, A.; Altman, D.G.; Tetzlaff, J.; Mulrow, C.; Gøtzsche, P.C.; Ioannidis, J.P.; Clarke, M.; Devereaux, P.J.; Kleijnen, J.; Moher, D. The PRISMA statement for reporting systematic reviews and meta-analyses of studies that evaluate health care interventions: Explanation and elaboration. *BMJ* **2009**, *339*, b2700, doi:10.1136/bmj.b2700.
3. Almazyad, A.; Li, C.C.; Tapia, R.O.C.; Robertson, J.P.; Collette, D.; Woo, S.B. Primordial odontogenic tumour: Report of two cases. *Histopathology* **2018**, *72*, 1221–1227, doi:10.1111/his.13488.
4. Bomfim, B.B; Prado, R.; Sampaio, R.K.; Conde, D.C.; de Andrade, B.A.B.; Agostini, M.; Romañach, M.J. Primordial Odontogenic Tumor: Report of a New Case and Literature Review. *Head Neck Pathol.* **2019**, *13*, 125–130, doi:10.1007/s12105-018-0913-7.

 © 2019 by the authors. Licensee MDPI, Basel, Switzerland. This article is an open access article distributed under the terms and conditions of the Creative Commons Attribution (CC BY) license (http://creativecommons.org/licenses/by/4.0/).

Extended Abstract

Serum and Salivary BP180 NC 16a Enzyme-Linked Immunosorbent Assay in Mucous Membrane Pemphigoid. Analysis of a Cohort of 25 Patients [†]

Fedora Della Vella [1,*], Sara Galleggiante [1], Claudia Laudadio [1], Maria Contaldo [2], Dario Di Stasio [2], Marilina Tampoia [1] and Massimo Petruzzi [1]

1. Interdisciplinary Department of Medicine, University of Bari "Aldo Moro", 70124 Bari, Italy; sara.galleggiante@libero.it (S.G.); c.lauda@hotmail.it (C.L.); mtampoia@libero.it (M.T.); massimo.petruzzi@uniba.it (M.P.)
2. Multidisciplinary Department of Medical-Surgical and Dental Specialties, University of Campania "Luigi Vanvitelli", 80138 Naples, Italy; maria.contaldo@gmail.com (M.C.); dario.distasio@unicampania.it (D.D.S.)
* Correspondence: dellavellaf@gmail.com; Tel.: +39-3498407957
† Presented at the XV National and III International Congress of the Italian Society of Oral Pathology and Medicine (SIPMO), Bari, Italy, 17–19 October 2019.

Published: 12 December 2019

1. Introduction

Mucous membrane pemphigoid (MMP) is a rare autoimmune blistering disease, affecting one or more mucosae. The oral cavity is generally involved, often in form of desquamative gingivitis (Figure 1).

Immunofluorescence (direct and indirect), histopathology (Figure 2), together with sera autoantibodies dosage by enzyme-linked immunosorbent assay (ELISA) are currently used to diagnose this disease. Saliva is currently employed for diagnostic purposes in multiple fields, providing a wide pool of biomarkers, and representing a not-invasive method suitable for screening and follow-up procedures [1,2].

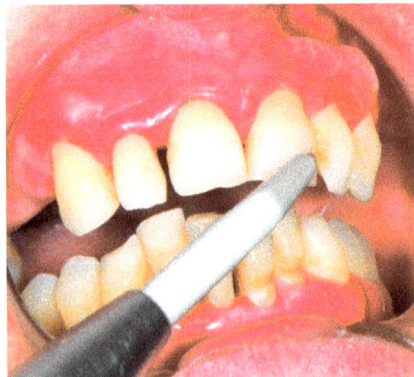

Figure 1. Desquamative gingivitis: common clinical oral manifestation of mucous membrane pemphigoid.

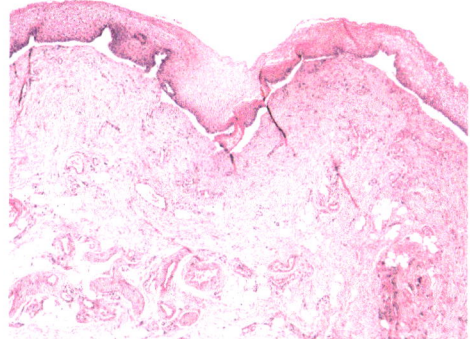

Figure 2. Histopathological picture of MMP (4× magnification).

2. Aim

The aims of the present study were to analyze the efficacy of ELISA in the detection of anti BP180 NC 16a antibodies, comparing salivary and sera specimens, in patients affected by MMP; and to evaluate the correlation between the severity of the clinical manifestations and the anti BP180 titers detected.

3. Materials & Method

Patients referred to Oral Medicine unit of Policlinico di Bari, from January 2011 to December 2016, with a histopathological diagnosis of MMP confirmed by direct immunofluorescence (DIF) test were enrolled. Every patient underwent a serum and a saliva sample collection at the beginning of the diagnostic process. BP180 titers were assessed using commercially available ELISA kit (cut off: 9 U/mL). Clinical Harman's score was recorded for every patient. Sensitivity, specificity, negative predictive value (NPV), positive predictive value (PPV) and accuracy of ELISA performed on salivary samples were calculated in comparison to sera specimens. The concordance between the two tests was evaluated using the Cohen's kappa coefficient (k). The correlation between clinical Harman's scores and ELISA titers was evaluated using t-Student test (significative values with $p < 0.05$).

4. Results

Twenty-five patients were enrolled, 19 females and 6 males (F:M = 3.16), mean age 63.5 (SD ± 10.9). Patients data are resumed in Table 1. Sensitivity and specificity values were 83.3% and 53.8% respectively, NPV was 77.8% and PPV was 62.5%, accuracy resulted with a value of 68%. The Cohen's k value was 0.367 (concordance: "fair"). The association between the clinical scores and both the salivary and sera anti-BP180 titers resulted to be statistically significant ($p < 0.05$).

Table 1. Enrolled patients' demographic data, anti-BP 180 sera and saliva titers detected using ELISA, and Harman's scores. The positive values (>9 U/mL) are bolded.

Patient	Sex	Age	Anti-BP180 Serum Titer	Anti-BP180 Saliva Titer	Herman Score
1	F	66	6.6	4.5	2
2	F	82	**43.2**	**26**	3
3	F	67	**13.6**	**19.7**	3
4	M	50	**28.8**	**44.3**	1
5	M	59	**30**	**9.1**	1
6	F	46	3	**11.2**	1
7	M	45	0.8	2.9	1
8	F	65	2.3	**40.9**	2
9	F	82	**26.3**	**11.2**	1
10	F	62	3.8	3.3	3
11	F	73	5	**42.1**	2

12	F	59	8.2	0.5	1
13	F	78	19.4	66.1	3
14	F	56	1.1	2.4	1
15	M	74	6.5	7.2	3
16	F	59	14.3	17.3	1
17	F	48	16.4	2.5	3
18	M	61	12.7	60	1
19	F	74	2	10	3
20	F	77	27.1	13	2
21	F	69	8.3	31.8	1
22	F	65	12.7	46.2	2
23	F	66	0.3	11.7	2
24	F	58	34.8	7.4	2
25	M	47	3.9	7.4	2

5. Discussion

From this study a good sensitivity of ELISA performed on saliva emerged, while specificity resulted lower when compared to sera data. Nowadays, few studies about ELISA effectiveness in patients affected by pemphigoid are available, and even less about muco-membranous forms. The concordance between the two samples used resulted to be fair, suggesting the chance to employ saliva as a not-invasive tool to help the diagnostic process and to track the course of the pathology [1–4]. In fact, this study showed a statistical correlation between clinical conditions and antibodies titers. A greater reliability of ELISA could be achieved extending the range of dosable antigens of the commercial kits, and using crevicular fluid samples instead of whole saliva, providing higher levels of antibodies.

Conflicts of Interest: The authors have no conflict of interests to declare.

References

1. Koopai, M.; Mortazavi, H.; Khatami, A.; Khodashenas, Z. Salivary and Serum Anti-Desmoglein 1 and 3 ELISA and Indirect Immunofluorescence in Pemphigus Vulgaris: Correlations with Serum ELISA, Indirect Immunofluorescence and Disease Severity. *Acta Dermatovenerol. Croat.* **2018**, *26*, 91–99.
2. Esmaili, N.; Mortazavi, H.; Kamyab-Hesari, K.; Aghazadeh, N.; Daneshpazhooh, M.; Khani, S.; Chams-Davatchi, C. Diagnostic accuracy of BP180 NC16a and BP230-C3 ELISA in serum and saliva of patients with bullous pemphigoid. *Clin. Exp. Dermatol.* **2015**, *40*, 324–330, doi:10.1111/ced.12510AA.
3. Andreadis, D.; Lorenzini, G.; Drakoulakos, D.; Belazi, M.; Mihailidou, E.; Velkos, G.; Mourellou-Tsatsou, O.; Antoniades, D. Detection of pemphigus desmoglein 1 and desmoglein 3 autoantibodies and pemphigoid BP180 autoantibodies in saliva and comparison with serum values. *Eur. J. Oral Sci.* **2006**, *114*, 374–380, doi:10.1111/j.1600-0722.2006.00394.x.
4. Ali, S.; Kelly, C.; Challacombe, S.J.; Donaldson, A.N.; Dart, J.K.; Gleeson, M.; MMP Study Group 2009–14; Setterfield, J.F. Salivary IgA and IgG antibodies to bullous pemphigoid 180 noncollagenous domain 16a as diagnostic biomarkers in mucous membrane pemphigoid. *Br. J. Dermatol.* **2016**, *174*, 1022–1029, doi:10.1111/bjd.14351.

© 2019 by the authors. Licensee MDPI, Basel, Switzerland. This article is an open access article distributed under the terms and conditions of the Creative Commons Attribution (CC BY) license (http://creativecommons.org/licenses/by/4.0/).

Extended Abstract

Application of Ozone Therapy in the Conservative Surgical Treatment of Osteonecrosis of the Jaw: Preliminary Results [†]

Rodolfo Mauceri [1,*], Anna Di Grigoli [1], Michele Giuliani [2], Marco Mascitti [3], Carmine Del Gaizo [1] and Olga Di Fede [1]

1. Department of Surgical, Oncological and Oral Sciences (Di.Chir.On.S.), University of Palermo, 90127 Palermo, Italy; anna.digrigoli@community.unipa.it (A.D.G.); delgaizocarmine@gmail.com (C.D.G.); odifede@odonto.unipa.it (O.D.F.)
2. Department of Clinical and Experimental Medicine, University of Foggia, 71122 Foggia, Italy; michele.giuliani@unifg.it
3. Department of Clinical Specialistic and Dental Sciences, Marche Polytechnic University, 60121 Ancona, Italy; marcomascitti86@hotmail.it
* Correspondence: rodolfo.mauceri@unipa.it
† Presented at the XV National and III International Congress of the Italian Society of Oral Pathology and Medicine (SIPMO), Bari, Italy, 17–19 October 2019.

Published: 11 December 2019

1. Objectives

The main goals of the management of osteonecrosis of the jaw (ONJ) are to slow the progression of the disease and, when it is achievable, to remove all the necrotic bone promoting the tissues' healing. In particular, the gold standard is represented by the surgical procedures (conservative or invasive) [1].

Recently, the use of medical ozone is increasingly applied in oral surgery, due to is valuable features, thanks to antimicrobial effect, regenerative and angiogenic activities [2].

This study aimed to evaluate the efficacy and safety of ozone application in the conservative surgical treatment of ONJ.

2. Materials and Methods

Twenty-three patients have been referred to our Sector of Oral Medicine (UNIPA) for ONJ treatment and have been enrolled in this study.

All patients have been staged, according to SICMF-SIPMO clinical and radiological staging system.

After informed consent, the PROMaF protocol [3] has been modified, adding insufflation/injection of ozone:

1. Antibiotic prophylaxis and one-minute mouthrinse with 0.2% Chlorhexidine from the day before and for six days after the surgical procedure.
2. Local anesthesia achieved using 3% mepivacaine hydrochloride without adrenaline
3. Elevation of a full-thickness mucoperiosteal flap
4. Curettage of the necrotic bone
5. Insufflation inside the bone defect (15 mL dosage) (by pink Venocat cannula 20Gx1.1/4″/1.10 × 32 mm) and injection around its edges (15 mL dosage) (by 26Gx½″ needle—0.45 × 13 mm) of an oxygen-ozone mixture (15γ concentration)
6. Tension-free suture

One week after, sutures have been removed; the follow-up has been expected at 15 days, 1, 3, 6 and 12 months after surgery; radiological evaluation has been carried out at 90 days, 6 and 12 months.

Local Ethics Committee (Azienda Ospedaliera Universitaria Policlinico Paolo Giaccone di Palermo) approved the study (record number N°1/2018).

3. Results

During the study period, twenty-three patients have been recruited, whose descriptive statistics are shown in Table 1.

At the most recent follow-up visit (mean follow-up 8.72 ± 5.2 mo), complete clinical healing has been observed in nine patients under bisphosphonates therapy (75%) and nine patients in treatment with denosumab or denosumab + bevacizumab (81%); in five patients, only clinical improvement has been observed. No recurrence signs have been showed in radiological findings.

Table 1. Patients' descriptive statistics.

Age (yrs)	65.4 ± 13
Male	7 (30%)
Female	16 (70%)
Cancer	17 (70%)
Non-Cancer	6 (30%)
Involved bone	
Maxilla	6 (30%)
Mandible	17 (70%)
ONJ stage *	
I A	4 (17%)
I B	5 (22%)
II A	4 (17%)
II B	7 (31%)
III B	3 (13%)
ONJ-related medications	
Bisphosphonates	12 (53%)
Denosumab	7 (30%)
Denosumab + Bevacizumab	4 (17%)
Time of administration of ONJ-related medications (mo)	
Cancer	21.3 ± 14.1
Non-Cancer	152 ± 77.1

* ONJ stage according to SICMF-SIPMO clinical and radiological staging system.

4. Conclusions

Although with the great limitation of these preliminary results, the authors suppose that ozone application may act as local regulators of wound healing, improving the results of ONJ surgical treatment.

Conflicts of Interest: The authors declare no conflict of interest.

References

1. Di Fede, O.; Mauceri, R.; Panzarella, V.; Maniscalco, L.; Bedogni, A.; Licata, M.E.; Albanese, A.; Toia, F.; Cumbo, E.M.G.; Mazzola, G.; et al. Conservative Surgical Treatment of Bisphosphonate-Related Osteonecrosis of the Jaw with Er, Cr: YSGG Laser and Platelet-Rich Plasma: A Longitudinal Study. *BioMed Res. Int.* **2018**, *2018*, 3982540, doi:10.1155/2018/3982540.
2. Ripamonti, C.I.; Maniezzo, M.; Boldini, S.; Pessi, M.A.; Mariani, L.; Cislaghi, E. Efficacy and tolerability of medical ozone gas insufflations in patients with osteonecrosis of the jaw treated with bisphosphonates—Preliminary data: Medical ozone gas insufflation in treating ONJ lesions. *J. Bone Oncol.* **2012**, *1*, 81–87, doi:10.1016/j.jbo.2012.08.001.

3. PROMaF Protocol: Prevention and Research on Medication—Related Osteonecrosis of the Jaws, 2014. Available online: http://www.policlinico.pa.it/portal/pdf/promaf/1.%20Cosa%20è%20PROMaF%20__ad%20uso%20degli%20operatori%20sanitari.pdf (accessed on 1 December 2014).

© 2019 by the authors. Licensee MDPI, Basel, Switzerland. This article is an open access article distributed under the terms and conditions of the Creative Commons Attribution (CC BY) license (http://creativecommons.org/licenses/by/4.0/).

Extended Abstract

Oral Oncology and Oral Medicine Fellowship for the General Dentist [†]

Paolo Junior Fantozzi [1,2,*], Michael Monopoli [3] and Alessandro Villa [1,2]

1. Department of Oral Medicine, Infection and Immunity, Harvard School of Dental Medicine, Boston, MA 02115, USA; avilla@bwh.harvard.edu
2. Division of Oral Medicine and Dentistry, Brigham and Women's Hospital, Boston, MA 02115, USA
3. DentaQuest Foundation, Boston, MA 02129, USA; Mike.Monopoli@dentaquest.com
* Correspondence: paolojfantozzi@gmail.com; Tel.: +39-3332-059-369
† Presented at the XV National and III International Congress of the Italian Society of Oral Pathology and Medicine (SIPMO), Bari, Italy, 17–19 October 2019.

Published: 11 December 2019

1. Aim or Purpose

Cancer patients and survivors are increasing [1] and may present with unique oral health needs. The majority of dentists do not have adequate education [2], training and experience [3] in the management of toxicities secondary to cancer therapy [4]. As such, we implemented a new oral oncology and oral medicine fellowship to meet these needs. The goal of this study was to determine the number of cancer patient visits, patient demographic information, and diagnoses in a single center training program.

2. Materials and Methods

A retrospective electronic medical record review was conducted from October 2018 till January 2019 for all patients seen by the fellow while rotating through the medical oncology departments at Dana Farber Cancer Institute. Patients were evaluated by the medical oncologist and the fellow at each visit. Data were tabulated into an electronic spreadsheet and descriptive statistics were used.

3. Results

586 patients (349 females, 59.6%) with a median age of 62 (range: 25-88) were seen. Of all patients seen, 402 (68.6%) had cancer. The most common cancer diagnoses included hematologic (179; 44.5%), head and neck (77; 18.6%), breast (73; 18.1%), thoracic (22; 5.5%) and bone malignancies (17; 4.2%).Oral complications were present in 96 (23.8%) patients and included dry mouth (60; 14.9%), ulcers (49; 12.2%), and chronic GHVD (25; 6.2%).Over the year, 122 procedures were performed (which 67.2% represented surgical extractions and biopsies) and 196 prescriptions for medications were done. Patients affected by head and neck cancer (35/77; 45.4%) and hematologic malignancies (55/179; 30.7%) had the highest risk of presenting with oral toxicities.

4. Conclusions

Oral complications from cancer therapy are common and may require a multidisciplinary approach. Adequately trained dentists are a valuable resource for the community to provide safe and effective oral care for oncology patients.

Conflicts of Interest: The authors declare no conflict of interest. The funding sponsors had no role in the design of the study; in the collection, analyses, or interpretation of data; in the writing of the manuscript, and in the decision to publish the results.

References

1. Shapiro, C.L. Cancer Survivorship. *N. Engl. J. Med.* **2018**, *379*, 2438–2450. doi:10.1056/NEJMra1712502
2. Cannick, G.F.; Horowitz, A.M.; Drury, T.F.; Reed, S.G.; Day, T.A. Assessing oral cancer knowledge among dental students in South Carolina. *J. Am. Dent. Assoc.* **2005**, *136*, 373–378. doi:10.14219/jada.archive.2005.0180.
3. Carter, L.M.; Ogden, G.R. Oral cancer awareness of general medical and general dental practitioners. *Bdj* **2007**, *203*, E10. doi:10.1038/bdj.2007.630.
4. Dyer, G.; Brice, L.; Schifter, M.; Gilroy, N.; Kabir, M.; Hertzberg, M.; Greenwood, M.; Larsen, S.R.; Moore, J.; Gottlieb, D.; et al. Oral health and dental morbidity in long-term allogeneic blood and marrow transplant survivors in Australia. *Aust. Dent. J.* **2018**, *63*, 312–319. doi:10.1111/adj.12627

© 2019 by the authors. Licensee MDPI, Basel, Switzerland. This article is an open access article distributed under the terms and conditions of the Creative Commons Attribution (CC BY) license (http://creativecommons.org/licenses/by/4.0/).

Extended Abstract

Use of Optical Coherence Tomography in a Patient with Erosive Oral Lichen Planus Treated with Low-Level Laser Therapy. Preliminary Findings [†]

Alessio Gambino [1,*], Marco Cabras [1], Adriana Cafaro [1], Paolo Giacomo Arduino [1], Paola Carcieri [1], Davide Conrotto [1], Mario Carbone [1], Luigi Chiusa [2], Adam Strange [3], Colin Hopper [3] and Roberto Broccoletti [1]

1. Department of Surgical Sciences, Oral Medicine Section, CIR-Dental School, University of Turin, 10126 Turin, Italy; cabrasmarco300@gmail.com (M.C.); adri.cafaro@gmail.com (A.C.); paologiacomo.arduino@unito.it (P.G.A.); carcieri.paola@libero.it (P.C.); davide.conrotto@gmail.com (D.C.); mario_carbone@libero.it (M.C.); roberto.broccoletti@unito.it (R.B.)
2. Department of Biomedical Sciences and Human Oncology, University of Turin, 10126 Turin, Italy; lchiusa@gmail.com
3. Department of Clinical Research, UCL Eastman Dental Institute, London WC1E 6DG, UK; adam.strange.13@ucl.ac.uk (A.S.); c.hopper@ucl.ac.uk (C.H.)
* Correspondence: alessio.gambino@unito.it; Tel.: +39-011-633-15-22
† Presented at the XV National and III International Congress of the Italian Society of Oral Pathology and Medicine (SIPMO), Bari, Italy, 17–19 October 2019.

Published: 11 December 2019

1. Introduction

Clinical studies have demonstrated the effectiveness of low-level laser therapy (LLLT) in patients with unresponsive oral lichen planus (OLP). OCT can reveal, in real time, the architecture of epithelial and sub-epithelial tissues and surrounding structures [1]. Aim of the present work was to assess the in-vivo changes of oral mucosa before and after LLLT treatment in a patient affected by erosive OLP.

2. Methods

OCT: a recent variant of a commercial frequency domain OCT dermatological instrument (SS-OCT, VivoSight®, version 2.0, Kent, UK) was deployed with a novel probe manufactured specifically for oral cavity. Length of probe = 124 mm; probe shaft diameter = 15 mm. Field of view = 6 mm². Dynamic scans—duration = 30 s, 120 frames, depth = 6 mm – allowing hyporeflectiveness to be displayed as red area, were deployed. OCT scans were carried out before biopsy, and before/after each LLLT session.

LLLT: 980 nm diode laser (Raffaello DMT Italy) was kept perpendicularly at 2 mm from the area of irradiation, with the following parameters: application time for each point = 16 s; total energy = 4 J; output power = 250 mW; power density = 500 mW/cm²; energy density = 8 J/cm²; spot size = 0.5 cm². A "spot" technique with slight overlapping was performed.

Patient: 74 years-old female, affected by histologically confirmed OLP, undergoing one weekly session of LLLT until complete clinical healing, for an atrophic-erosive cheek lesion of 1 cm², unresponsive to topical steroids.

3. Results

Before biopsy OCT revealed a wide red hyporeflective area beneath and above the basement membrane (BM, green line in Figure 1). Comparison between biopsy specimen and OCT scan suggested that hyporeflectiveness beneath BM might be ascribed to an uprising of the blood flow in the inflamed connective tissue, whereas above BM, hyporeflectiveness might be attributed to the edema within the epithelial layers. Immediately after the first LLLT session, despite any detectable clinical modifications, OCT revealed a decrease of the hyporeflectiveness of the upper third of the epithelium (Figure 2). After the second LLLT session, OCT showed a supplementary decrease of the red area beneath and above BM. Finally, after the third LLLT session, clinical healing was accomplished, with OCT showing a homogeneous red zone limited to the area beneath BM, whereas epithelium regained a usual hyper-reflective gray pattern intertwined with red spikes.

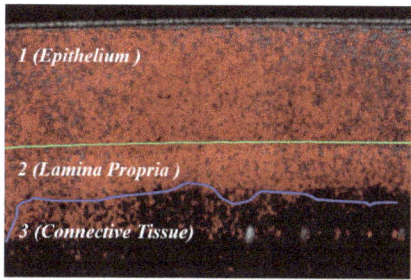

Figure 1. OCT dynamic scan of the cheek lesion before biopsy, showing widespread red hyporeflective areas both in Epithelium and Lamina Propria.

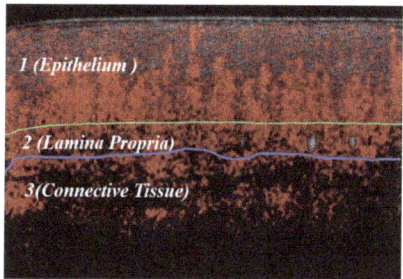

Figure 2. OCT dynamic scan of the cheek lesion immediately after the first LLLT session, showing a decrease of the hyporeflectiveness of the upper third of the epithelium.

To our knowledge, this is the first evidence of application of OCT in oral medicine as an additional tool to assess LLLT reliability for OLP. Interestingly, OCT seemed to reveal modifications of the epithelial ultrastructure before the appearance of visible clinical changes. Larger samples of OLP patients should be tested to assess the validity of these preliminary results.

Acknowledgments: Thanks to Vivosight® for leasing of the OCT.

Conflicts of Interest: The Authors declare no conflict of interest.

Reference

1. Standish, B.A.; Lee, K.K.; Mariampillai, A.; Munce, N.R.; Leung, M.K.; Yang, V.X.; Vitkin, I.A. In vivo endoscopic multi-beam optical coherence tomography. *Phys. Med. Biol.* **2010**, *55*, 615.

 © 2019 by the authors. Licensee MDPI, Basel, Switzerland. This article is an open access article distributed under the terms and conditions of the Creative Commons Attribution (CC BY) license (http://creativecommons.org/licenses/by/4.0/).

Extended Abstract
Oral Manifestations of Psoriasis: A Systematic Review [†]

Gioele Gioco [1,*], Romeo Patini [1], Giuseppe Troiano [2], Alessia Di Petrillo [1], Patrizia Gallenzi [1], Luisa Limongelli [3] and Carlo Lajolo [1]

[1] Fondazione Policlinico Universitario A. Gemelli IRCSS, Università Cattolica del Sacro Cuore, Head and Neck Department, Largo A. Gemelli, 8, 00168 Rome, Italy; romeo.patini@unicatt.it (R.P.); alessiadipetrillo@hotmail.it (A.D.P.); patrizia.gallenzi@unicatt.it (P.G.); carlo.lajolo@unicatt.it (C.L.)
[2] Department of Clinical and Experimental Medicine, University of Foggia, Via Rovelli 50, 71122 Foggia, Italy; giuseppe.troiano@unifg.it
[3] Department of Interdisciplinary Medicine, Complex Operating Unit of Odontostomatology, Aldo Moro University, Piazza Giulio Cesare 11, 70124 Bari, Italy; lululimongelli@gmail.com
* Correspondence: gioele.gioco@hotmail.it; Tel.: +39-3288-486-585
[†] Presented at the XV National and III International Congress of the Italian Society of Oral Pathology and Medicine (SIPMO), Bari, Italy, 17–19 October 2019.

Published: 11 December 2019

1. Introduction

Psoriasis is a systemic, immune-mediated, inflammatory skin disease of unknown aetiology [1]. It affects 2–4% of the population in western countries and it is characterized by circumscribed, circular, red papules or plaques with a grey or silvery-white, dry scale.

Although many studies reported possible association between psoriasis and several oral conditions [2], some questions still remain unsolved.

The aim of this review was to analyze all the existing literature on oral manifestations in psoriatic patients and to determine their prevalence compared with healthy subjects.

2. Methods

PubMed, Scopus, and Web of Science were used as search engines. Only observational, full-length, English language studies were investigated. PRISMA checklist was used as guideline for this review. PROSPERO registration code is CRD42019127178.

3. Results

Among 3580 records screened, only 40 were included in this review. Meta-analysis was performed on the included case-control studies and mean prevalence and pooled odds ratio were calculated.

Among the included studies, the mean prevalence of geographic tongue was 11.80% in case group vs. 6.59% in control group (OR= 2.81, 95% CI 2.2–3.6); the mean prevalence of fissured tongue was 29.96% in case group vs. 13.90% in control group (OR = 2.93, 95% CI 2.5–3.5); the mean prevalence of periodontitis was 26.29% in case group vs. 11.48% in control group (OR = 3.5, 95% CI 2.49–4.23); the mean prevalence of candidiasis was 6.67% in case group *vs* 1.04% in control group (OR = 3.17, 95% CI 1.8–5.7); the mean prevalence of *Candida* spp. was 45.81% in case group vs. 20.53% in control group (OR = 3.79, 95% CI 2.3–6.4); the mean prevalence of TMD was 40.67% in case group *vs.* 16.96% in control group (OR = 2.94, 95% CI 1.1–7.9).

Due to dis-homogeneity across studies no quantitative analysis was performed on salivary biochemical composition. Nevertheless, psoriatic patients presented a lower mean concentration of salivary IgA, lysozyme and cortisol than controls, whereas IL-1β was higher.

4. Discussion

This systematic review revealed that geographic tongue, fissured tongue, periodontitis, candidiasis and TMD are more frequent in psoriatic patients than general population. Moreover, psoriatic patients showed biochemical alterations in the saliva composition compared to healthy subjects. Future research should be conducted to rule out the mechanism underlying these associations and to investigate the behavior of oral manifestation in psoriatic patients in relation to their severity and treatment response.

Conflicts of Interest: The authors declare no conflict of interest.

References

1. Yesudian, P.D.; Chalmers R.J.; Warren, R.B.; Griffiths, C.E. In search of oral psoriasis. *Arch. Dermatol. Res.* **2012**, *304*, 1–5. doi:10.1007/s00403-011-1175-3.
2. Zhu, J.F.; Kaminski, M.J.; Pulitzer, D.R.; Hu, J.; Thomas, H.F. Psoriasis: Pathophysiology and oral manifestations. *Oral Dis.* **1996**, *2*, 135–144.

 © 2019 by the authors. Licensee MDPI, Basel, Switzerland. This article is an open access article distributed under the terms and conditions of the Creative Commons Attribution (CC BY) license (http://creativecommons.org/licenses/by/4.0/).

Extended Abstract

Teeth Extractions in Subjects Undergoing Radiotherapy for Head and Neck Cancers: A Systematic Review on the Clinical Protocols for Preventing Osteoradionecrosis (ORN). Extractions after Radiotherapy (Part 2) †

Gioele Gioco [1,*], Cosimo Rupe [1], Giuseppe Troiano [2], Michele Giuliani [2], Massimo Petruzzi [3] and Carlo Lajolo [1]

[1] Head and Neck Department, "Fondazione Policlinico Universitario A. Gemelli–IRCCS", School of Dentistry, Università Cattolica del Sacro Cuore, 00168 Rome, Italy; cosimorupe@gmail.com (C.R.); carlo.lajolo@unicatt.it (C.L.)
[2] Department of Clinical and Experimental Medicine, University of Foggia, 71122 Foggia, Italy; giuseppe.troiano@unifg.it (G.T.); michele.giuliani@unifg.it (M.G.)
[3] Interdisciplinary Department of Medicine, University "Aldo Moro" of Bari, 70121 Bari, Italy; massimo.petruzzi@uniba.it
* Correspondence: gioele.gioco@hotmail.it; Tel.: +39-3288486585
† Presented at the XV National and III International Congress of the Italian Society of Oral Pathology and Medicine (SIPMO), Bari, Italy, 17–19 October 2019.

Published: 11 December 2019

Osteoradionecrosis (ORN) of the jaws is the most severe side effect, and in some case life-threatening, of radiotherapy for head and neck cancer [1]. Tooth extractions during and after radiotherapy are the major risk factor of ORN onset, nevertheless the real ORN rate and its risk factors are still unclear. Recently, a systematic review [2] tried to better estimate the ORN rate following tooth extractions after RT, pointing out an overall rate of 7%, nevertheless the high number of variables involved in the onset of ORN makes the topic particularly complex and many information still lack to prevent the ORN onset after tooth extraction. The aim of this systematic review is to assess the incidence of ORN in those patients who underwent tooth extraction during and after radiotherapy for head and neck cancer and to identify any possible risk factors.

PRISMA protocol was used to evaluate and present the results. PubMed, Scopus, and Web of Science were used as search engines: only English full-length papers of clinical trial and observational studies both prospective and retrospective, published in peer-reviewed journals, were investigated. Inclusion criteria were: minimum sample size 10 patients who underwent tooth extractions during and after radiotherapy; 6 months mean follow-up after tooth extraction; no previous ORN at extraction site; the ORN diagnosis criteria must be clearly defined in the text; must be specified if ORN developed at extraction site or not. Cumulative meta-analysis was performed calculating the pooled proportion (PP) of the rate of ORN occurrence. Meta-analysis was performed at random effects model with the Der-Simonian Liard method. All the statistical analyses were performed with the software Open Meta-Analyst version 10. PROSPERO registration code is CRD42018079986.

Among 2020 records screened, only 9 were included in this review (Figure 1). Forty-one of 462 patients who underwent tooth extraction during and after radiotherapy developed ORN at extraction site, with an ORN incidence of 5.8% (95% Confidence of Interval = 2.5–9.4, $p < 0.001$, $I^2 = 8322\%$).

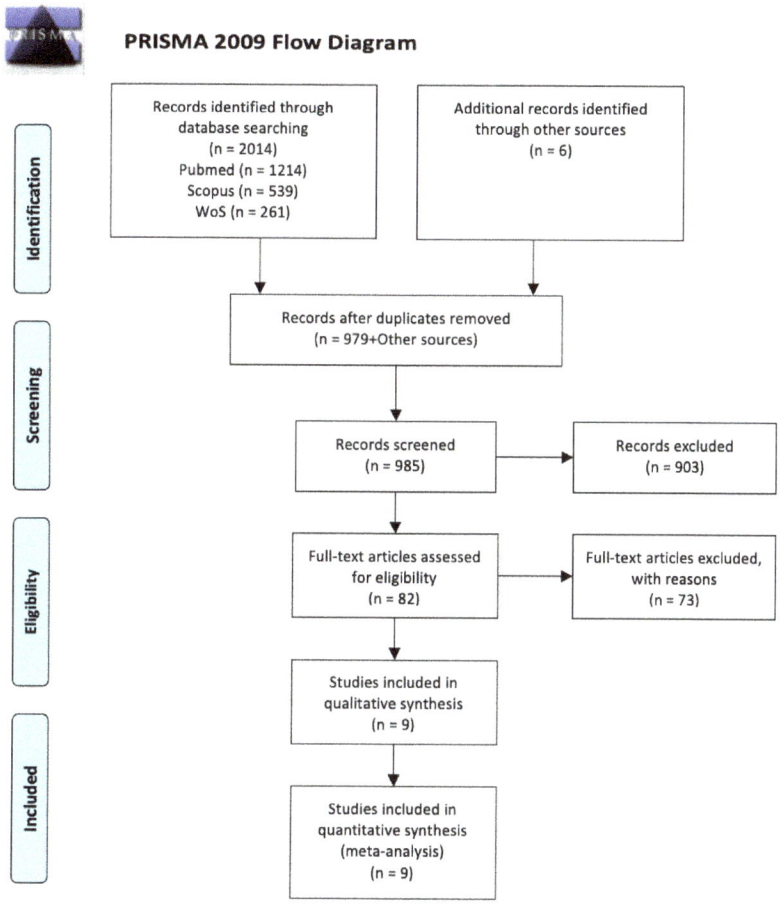

Figure 1. PRISMA flow-chart of the inclusion process.

This systematic review confirmed dental extraction after radiotherapy as the main risk factor of ORN and its incidence ranges between 6% and 11% (Figure 2). The considerable amount of missing data didn't allow us to identify other possible risk factors for ORN onset.

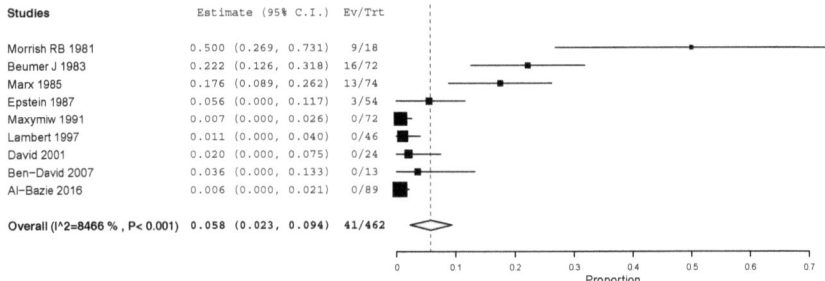

Figure 2. Pooled data and box plot stratifying for pathological criteria.

Although the most recently performed studies highlight a considerable decreasing of ORN incidence, the results confirm the recommendation to perform a dental evaluation and to remove oral foci before the beginning of radiotherapy.

Conflicts of Interest: The authors declare no conflict of interest.

References

1. Nadella, K.; Kodali, R.; Guttikonda, L.; Jonnalagadda, A. Osteoradionecrosis of the Jaws: Clinico-Therapeutic Management: A Literature Review and Update. *J. Maxillofac. Oral Surg.* **2015**, *14*, 891–901.
2. Nabil, S.; Samman, N. Incidence and prevention of osteoradionecrosis after dental extraction in irradiated patients: A systematic review. *Int. J. Oral Maxillofac. Surg.* **2011**, *40*, 229–243.

 © 2019 by the authors. Licensee MDPI, Basel, Switzerland. This article is an open access article distributed under the terms and conditions of the Creative Commons Attribution (CC BY) license (http://creativecommons.org/licenses/by/4.0/).

Extended Abstract

Clinical Validation of 13-Gene DNA Methylation Analysis from Oral Brushing: A Non Invasive Sampling Procedure for Early Detection of Oral Squamous Cell Carcinoma. A Multicentric Study [†]

Davide B. Gissi [1,*], Umberto Romeo [2], Gianluca Tenore [2], Monica Pentenero [3], Giuseppina Campisi [4], Rodolfo Mauceri [4], Giuseppe Colella [5], Roberto De Luca [5], Rosario Serpico [6], Dario Di Stasio [6], Giacomo Oteri [7], Paolo Vescovi [8], Michele D. Mignogna [9], Noemi Coppola [9], Andrea Santarelli [10], Luca Morandi [11] and Lucio Montebugnoli [1]

[1] Department of Biomedical and Neuromotor Sciences, Section of Oral Sciences, University of Bologna, 40159 Bologna, Italy; lucio.montebugnoli@unibo.it
[2] Department of Oral and Maxillofacial Sciences, "Sapienza" University of Rome, 00100 Rome, Italy; umberto.romeo@uniroma1.it (U.R.); gianluca.tenore@uniroma1.it (G.T.)
[3] Oral Medicine and Oral Oncology Unit, Department of Oncology, University of Turin, 10043 Orbassano, Italy; monica.pentenero@unito.it
[4] Department of Surgical, Oncological and Stomatological Disciplines, Sector of Oral Medicine, University of Palermo, 90127 Palermo, Italy; giuseppina.campisi@unipa.it (G.C.); rodolfo.mauceri@unipa.it (R.M.)
[5] Multidisciplinary department of Medical, Surgical and Dental Specialty, Maxillofacial Surgery Unit, University of Campania "Luigi Vanvitelli", 80100 Naples, Italy; giuseppe.colella@unicampania.it (G.C.); robertodeluca89@yahoo.it (R.D.L.)
[6] Multidisciplinary Department of Medical-Surgical and Dental Specialties, University of Campania "Luigi Vanvitelli", 80100 Naples, Italy; rosario.serpico@unicampania.it (R.S.); dario.distasio@unicampania.it (D.D.S.)
[7] Department of Biomedical and Dental Sciences and Morphofunctional Imaging, University of Messina, University Hospital "Gaetano Martino", 98124 Messina, Italy; oterig@unime.it
[8] Department of Medicine and Surgery, Oral Medicine and Laser Surgery Unit, University of Parma, 43100 Parma, Italy; paolo.vescovi@unipr.it
[9] Head & Neck Clinical Section, Department of Neuroscience, Reproductive and Odontostomatological Sciences, Federico II University of Naples, 80138 Naples, Italy; mignogna@unina.it (M.D.M.); noemi.coppola91@gmail.com (N.C.)
[10] Department of Clinical Sciences and Stomatology, Marche Polytechnic University, 60121 Ancona, Italy; andrea.santarelli@staff.unipvm.it
[11] Department of Biomedical and Neuromotor Sciences, Functional MR Unit, IRCCS Istituto delle Scienze Neurologiche di Bologna, University of Bologna, 40139 Bologna, Italy; luca.morandi2@unibo.it
* Correspondence: davide.gissi@unibo.it; Tel.: +39-0512088123
† Presented at the XV National and III International Congress of the Italian Society of Oral Pathology and Medicine (SIPMO), Bari, Italy, 17–19 October 2019.

Published: 11 December 2019

1. Introduction

In a recent study our research group described a non-invasive sampling procedure based on DNA methylation analysis of a set of 13 genes with a high level of accuracy (sensitivity 96.6%, specificity 100%) in the detection of squamous cell carcinoma of the oral cavity (OSCC) [1].

The purpose of the present study was to test the diagnostic performance of this non invasive sampling procedure in an italian multicentric study.

2. Materials and Methods

Oral brushing specimens were collected in ten different italian units of oral medicine. Each oral medicine unit collected blindly 10 brushing specimens from patients affected by OSCC and an equal number of age and sex-matched healthy controls. 13-gene DNA methylation analysis was performed and each sample was considered positive or negative in relation to a predefined cut-off value.

3. Results

181 out of 200 planned specimens were analyzed. DNA could not be amplified in 4 cases (2.2%). 86/93 (92.5%) specimens derived from OSCC patients were detected as positive and 70/84 (83.3%) specimens derived from healthy donors showed a negative score.

4. Conclusions

Data from multicentric study confirmed a high level of sensitivity of our procedure whereas level of specificity is slightly lower if compared to our previous study. These data suggest that our procedure may be proposed as a first level diagnostic test with the aim to avoid a diagnostic delay in Oral Squamous Cell Carcinoma.

Conflicts of Interest: As a possible conflict of interest, L. Morandi and D.B.G. submitted a patent (the applicant is the University of Bologna) in November 2016 to the National Institute of 398 Industrial Property; however, we believe that this is a natural step of translational research (bench-to-bedside) 399 and guarantee that the scientific results are true. The remaining authors declare that they have no competing 400 interest.

References

1. Morandi, L.; Gissi, D.; Tarsitano, A.; Asioli, S.; Gabusi, A.; Marchetti, C.; Montebugnoli, L.; Foschini, M.P. CpG location and methylation level are crucial factors for the early detection of oral squamous cell carcinoma in brushing samples using bisulfite sequencing of a 13-gene panel. *Clin. Epigenet.* **2017**, *9*, 85.

© 2019 by the authors. Licensee MDPI, Basel, Switzerland. This article is an open access article distributed under the terms and conditions of the Creative Commons Attribution (CC BY) license (http://creativecommons.org/licenses/by/4.0/).

Extended Abstract

Intraoral Ultrasound in the Evaluation of Depth of Invasion in OSCC. Preliminary Results [†]

Francesca Graniero [1,*], Leonardo D'Alessandro [1,*], Alessandra Montori [1,*], Federica Rocchetti [1,*], Vito Cantisani [2,*], Andrea Cassoni [1,*], Gianluca Tenore [1,*] and Umberto Romeo [1,*]

1. Department of Oral and Maxillofacial Sciences, Sapienza University of Rome, 00161 Rome, Italy
2. Department of Radiological, Oncological and Anatomo-Pathological Sciences, Sapienza University of Rome, 00161 Rome, Italy
* Correspondence: francesca.graniero@outlook.it (F.G.); dalessandro.1634449@gmail.com (L.D.); montorialessandra@gmail.com (A.M.); federica.rocchetti@uniroma1.it (F.R.); vito.cantisani@uniroma1.it (V.C.); andrea.cassoni@uniroma1.it (A.C.); gianluca.tenore@uniroma1.it (G.T.); umberto.romeo@uniroma1.it (U.R.)
† Presented at the XV National and III International Congress of the Italian Society of Oral Pathology and Medicine (SIPMO), Bari, Italy, 17–19 October 2019.

Published: 10 December 2019

Oral Squamous Cell Carcinoma (OSCC) shows an early tendency to lymphatic spread rather than hematogenous. The surgical treatment cannot be considered oncologically complete if the neck is not evaluated [1]. According to TNM staging system, the management of early stage (T1/2) or clinically node-negative, is still controversial (Figure 1). Several studies have shown that tumour thickness and depth can be considered the most important prognostic factors; Depth of Invasion (DOI) means the cancer growth extension into the tissue while thickness concerns the entire mass [1]. The exact depth cut-off has not yet been well defined. A preoperative investigation of tumour thickness and DOI would provide useful informations for targeting those patients who need neck treatment. To measure these factors are available Magnetic Resonance Imaging (MRI), Computed Tomography (CT) and Ultrasonography (US) [2]. The limitation of MRI and CT is that within a thickness less than 5 mm, it could be difficult to differentiate the tumour from the surrounding tissues. With the introduction of intraoral probes, US allows the direct evaluation of tumour and also it has advantages like harmless, radiation free, easy-to-use, non-invasive, unaffected by metal artefacts.

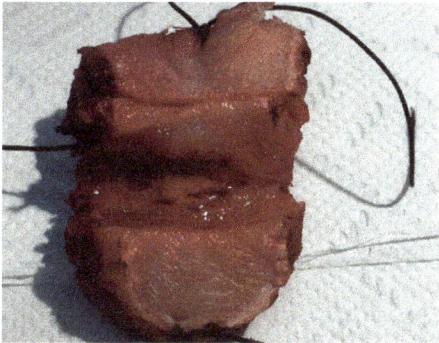

Figure 1. T1 tongue tumour.

The aim of this study was to compare tumour DOI by US with histological sections and to insert the US in the OSCC diagnostic flow-chart [1,2].

Twelve patients with histological diagnosis of OSCC T1 were undergone to ultrasound using an E-CUBE 15 EX scanner (Alpinion, Seoul, Korea) with a 8–17 MHz intraoral transducer like a toothbrush (Figure 2). For each patient has been performed an intra-operative and post-operative histological examination to establish tumour depth. Statistical analysis was made with SPSS 24 software (IBM, New York, NY, USA).

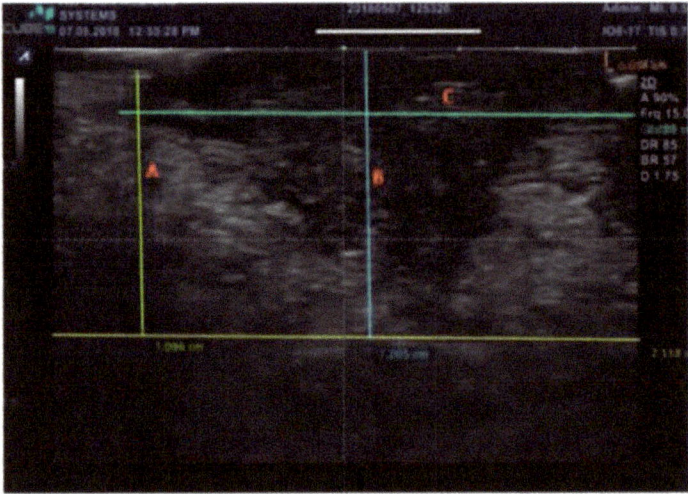

Figure 2. DOI (A), thickness (B), diameter (C).

By considering the presence of tumour infiltration, 90% sensitivity was found for intraoral ultrasound in comparison to histological evaluation. 9 true-positive, 2 false-positive and 1 false-negative occurred in our patients.

Using the Fisher Test, it was found that there was not a statistically difference between ultrasound DOI and histological DOI (chi-square = 0.218; p = 1000) Table 1.

Table 1. Comparison between ultrasound and hystological DOI.

		Infiltration		Total
		Infiltrating	Not INFILTRATING	
Method	Ultrasound	11	1	12
	Hystologic	10	2	12
Total		23	1	24

Although larger samples are needed, these preliminary results show that US is accurate to assess DOI level and it represents an useful and cost-effective device in the OSCC management [1,2].

Conflicts of Interest: The authors declare no conflict of interest.

References

1. Tarabichi, O.; Bulbul, M.G.; Kanumuri, V.V.; Faquin, W.C.; Juliano, A.F.; Cunnane, M.E.; Varvares, M.A. Utility of intraoral ultrasound in managing oral tongue squamous cell carcinoma: Systematic review. *Laryngoscope* **2019**, *129*, 662–670.
2. Angelelli, G.; Moschetta, M.; Limongelli, L.; Albergo, A.; Lacalendola, E.; Brindicci, F.; Favia, G.; Maiorano, E. Endocavitary sonography of early oral cavity malignant tumors. *Head Neck* **2017**, *39*, 1349–1356.

 © 2019 by the authors. Licensee MDPI, Basel, Switzerland. This article is an open access article distributed under the terms and conditions of the Creative Commons Attribution (CC BY) license (http://creativecommons.org/licenses/by/4.0/).

Extended Abstract

p63 Expression in Solitary and Syndromic Odontogenic Keratocysts: An Immunohistochemical Study [†]

Elisa Luconi [1,*], Lucrezia Togni [1], Giovanni Giannatempo [2], Vito Carlo Alberto Caponio [2], Marco Mascitti [1] and Andrea Santarelli [1]

1. Department of Clinical Specialistic and Dental Sciences, Marche Polytechnic University, 60126 Ancona, Italy; togni.lucrezia@gmail.com (L.T.); marcomascitti86@hotmail.it (M.M.); andrea.santarelli@staff.univpm.it (A.S.)
2. Department of Clinical and Experimental Medicine, University of Foggia, 71122 Foggia, Italy; edottor@libero.it (G.G.); vito_caponio.541096@unifg.it (V.C.A.C.)
* Correspondence: eliluconi@gmail.com; Tel.: +39-071-220-6226
† Presented at the XV National and III International Congress of the Italian Society of Oral Pathology and Medicine (SIPMO), Bari, Italy, 17–19 October 2019.

Published: 10 December 2019

In the last years, the classification of odontogenic cysts and tumors has been highly debated, especially regarding odontogenic keratocyst (OKC). Indeed OKC, previously defined as keratocistic odontogenic tumor, has been re-introduced into the cyst group due to the lack of sufficient evidence to justify its classification as a neoplasm [1,2]. OKC is characterized by a thin keratinizing stratified squamous epithelial lining and a non-inflammatory fibrous connective wall. OKC may occur in association with the nevoid basal cell carcinoma syndrome (NBCCS), characterized by an autosomal dominant gene mutation with variable and usually high penetrance and expressivity.

Due to the almost restricted expression of p63 in epithelial cells, its role in the regulation of proliferation and differentiation of craniofacial-structure, this study investigated immunohistochemical expression of p63 both in solitary OKC and NBCCS-related OKC.

10 solitary OKC as well as 5 NBCCS-related OKC were analysed by immunohistochemistry for expression of p63. 4-μm serial sections were incubated with the monoclonal anti-p63 antibody (clone 4A4, Dako Cytomation, Glostrup, Denmark) diluted 1:50.

A semi-quantitative assessment of p63 expression was performed: "−" (stained cells: 0–5%); "+" (stained cells: 6–30%); "++" (stained cells: 31–50%); "+++" (stained cells: >50%). Staining intensity was graded on a scale of a semi-quantitative scale: 0 (no detectable); 1 + (weak); 2 + (moderate); 3 + (strong); 4 + (very strong).

In NBCCS-related OKC 2 cases showed negligible p63 expression (0–5%), 2 cases showed a percentage of stained cells of 6–30%, and 1 case demonstrated a significant expression (31–50%). In all solitary OKC, the number stained cells were comprised between 0 and 5%. Regarding staining intensity, in NBCCS-related OKC the nuclear staining of p63 was intense (3+) in all epithelial layers. On the contrary, in solitary OKC, the intensity of staining of p63 was moderate (2+) and reported only in basal-parabasal layer.

In conclusion, these data suggest that that p63 diffuse and intense immunostaining could be useful to distinguish solitary OKC from NBCCS-related OKC.

Conflicts of Interest: The authors declare no conflict of interest.

References

1. El-Naggar, A.; Chan, J. *WHO Classification of Head and Neck Tumours*, 4th ed.; IARC: Lyon, France, 2017; pp. 235–237.
2. Lo Muzio, L.; Mascitti, M. Cystic lesions of the jaws: A retrospective clinicopathologic study of 2030 cases. *Oral Surg. Oral Med. Oral Pathol. Oral Radiol.* **2017**, *124*, 128–138, doi:10.1016/j.oooo.2017.04.006.

© 2019 by the authors. Licensee MDPI, Basel, Switzerland. This article is an open access article distributed under the terms and conditions of the Creative Commons Attribution (CC BY) license (http://creativecommons.org/licenses/by/4.0/).

Extended Abstract

IFI16 and Anti-IFI16 as Novel Biomarkers for Sjoegren's Syndrome: Preliminary Data †

Sonia Marino [1], Roberta Gualtierotti [2,*], Valeria Caneparo [3], Marco De Andrea [3,4], Marisa Gariglio [3], Pier Luigi Meroni [5], Eleonora Bossi [1] and Francesco Spadari [1]

[1] Department of Biomedical, Surgical and Dental Sciences-University of Milan, Maxillo-Facial and Odontostomatology Unit-Ospedale Policlinico, Fondazione IRCCS Ca' Granda, 20122 Milano, Italy; sonia.marino@studenti.unimi.it (S.M.); bossi.eleonora@virgilio.it (E.B.); francesco.spadari@unimi.it (F.S.)
[2] Department of Medical Biotechnology and Translational Medicine, University of Milan, 20129 Milan, Italy
[3] Intrinsic Immunity Unit, CAAD—Center for Translational Research on Autoimmune and Allergic Disease, University of Eastern Piedmont, 28100 Novara, Italy; valeria.caneparo@med.uniupo.it (V.C.); marco.deandrea@med.uniupo.it (M.D.A.); marisa.gariglio@med.uniupo.it (M.G.)
[4] Department of Public Health and Pediatric Sciences, University of Turin – Medical School, 10126 Turin, Italy
[5] Immunorheumatology Laboratory, Istituto Auxologico Italiano IRCCS, 20095 Milan, Italy; pierluigi.meroni@unimi.it
* Correspondence: roberta.gualtierotti@unimi.it; Tel.: +39-025-031
† Presented at the XV National and III International Congress of the Italian Society of Oral Pathology and Medicine (SIPMO), Bari, Italy, 17–19 October 2019.

Published: 11 December 2019

Sjoegren's syndrome (SS) is a chronic autoimmune disease characterized by sicca syndrome and systemic manifestations [1]. IFNγ-inducible protein-16 (IFI16) is a viral DNA sensor involved in infections and autoimmune diseases. In SS patients, IFI16 and anti-IFI16 antibodies can be detected in serum and salivary glands [2]. However, to date none of these findings were correlated with SS severity and disease activity.

IFI16 and anti-IFI16 in serum, minor salivary glands and saliva were evaluated together with clinical characteristics and EULAR SS disease activity index (ESSDAI) of SS patients.

Serum and tissue samples were analyzed as previously described [2], salivary anti-IFI16 IgG/IgA detection via ELISA based on horseradish peroxidase–conjugated rabbit anti-human IgG/IgA.

Table 1 describes patient characteristics. Patient (a), with moderate systemic SS (Figure 1a), showed increased IFI16 expression in inflammatory and epithelial cells not only in nuclei, but also in the cytosol, very high serum IFI16 (472 ng/mL), and serum anti-IFI16 IgG (116 U/mL). Patient (b), with localized SS (Figure 1b), showed moderately increased nuclear expression of IFI16 both in inflammatory and epithelial cells and no serum IFI16/anti-IFI16 IgG. In patient (c), with mild systemic SS (Figure 1c), we observed poor inflammatory infiltrate, although salivary anti-IFI16 IgA (10 U/mL) and serum anti-IFI16 IgG (147 U/mL) were found.

In saliva, IFI16 was not detected by Western blot. Recombinant IFI16 incubated at 37 °C was markedly degraded after 1h and completely degraded after 6 h in controls; completely degraded after 1h in patient (a), and more slowly degraded in patients (b) and (c).

Our results suggest that the expression and localization of IFI16 and anti-IFI16 may vary based on disease severity and activity of SS. Further experiments in a larger cohort of patients will allow us to better define the diagnostic and prognostic value of these biomarkers.

Table 1. Characteristics of patients.

Parameter	Patient (a)	Patient (b)	Patient (c)
ESSDAI	12	0	3
RF	+	-	-
Anti-SSA/Ro	+++	++	+
Anti-IFI16 IgG U/mL Serum	116 (+)	102 (-)	147 (+)
IFI16 ng/mL Serum	472 (++)	0.00 (-)	0.00 (-)
Anti-IFI16 IgA U/mL Saliva	4.7 (-)	5.4 (-)	10 (+)
Hypocomplementemia	+	-	+
Hypergammaglobulinemia	+	-	+

EULAR Sjoegren Syndrome Disease Activity Index, IFI16 IFNγ-inducible protein 16, RF rheumatoid factor, − negative, + positive, ++ high titre positivity, +++ very high titre positivity.

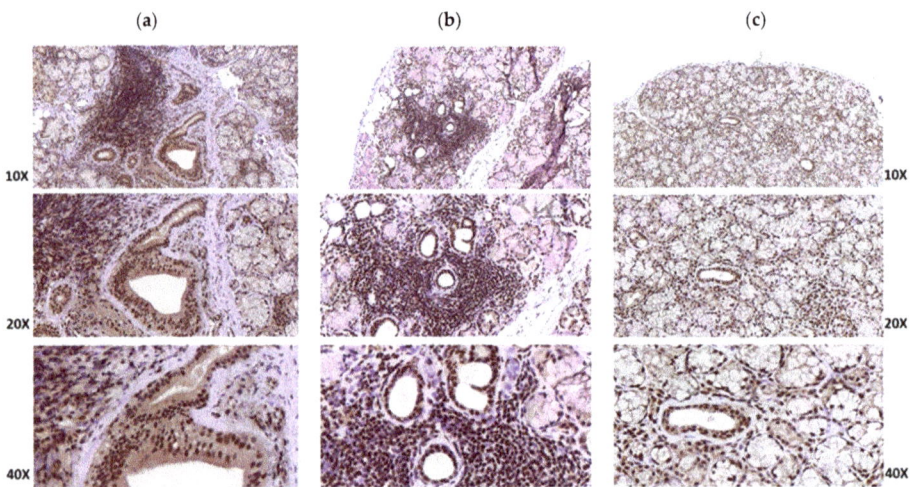

Figure 1. Immunohistochemistry for IFI16 expression in minor salivary glands in three patients. IFI16 staining is shown in brown and hematoxylin staining to highlight nuclei is in blue. (**a**) moderate systemic disease, (**b**) localized disease, (**c**) mild systemic disease.

Conflicts of Interest: The authors declare no conflict of interest.

References

1. Brito-Zerón, P.; Baldini, C.; Bootsma, H.; Bowman, S.J.; Jonsson, R.; Mariette, X.; Sivils, K.; Theander, E.; Tzioufas, A.; Ramos-Casals, M. Sjoegren's syndrome. *Nat. Rev. Dis. Primers.* **2016**, *2*, 16047.
2. Alunno, A.; Caneparo, V.; Carubbi, F.; Bistoni, O.; Caterbi, S.; Bartoloni, E.; Giacomelli, R.; Gariglio, M.; Landolfo, S.; Gerli, R. Interferon gamma-inducible protein 16 in primary Sjögren's syndrome: A novel player in disease pathogenesis? *Arthritis Res. Ther.* **2015**, *17*, 208.

 © 2019 by the authors. Licensee MDPI, Basel, Switzerland. This article is an open access article distributed under the terms and conditions of the Creative Commons Attribution (CC BY) license (http://creativecommons.org/licenses/by/4.0/).

Extended Abstract

Odontogenic Cysts: A 30-Year Retrospective Clinicopathological Study [†]

Marco Mascitti [1,*], Lucrezia Togni [1], Lorenzo Lo Muzio [2], Giuseppina Campisi [3], Federico Mazzoni [1] and Andrea Santarelli [1]

1. Department of Clinical Specialistic and Dental Sciences, Marche Polytechnic University, 60126 Ancona, Italy; togni.lucrezia@gmail.com (L.T.); edottor@libero.it (F.M.); andrea.santarelli@staff.univpm.it (A.S.)
2. Department of Clinical and Experimental Medicine, University of Foggia, 71122 Foggia, Italy; lorenzo.lomuzio@unifg.it
3. Department of Surgical, Oncological and Oral Sciences (DICHIRONS), University of Palermo, 90127 Palermo, Italy; campisi@odonto.unipa.it
* Correspondence: marcomascitti86@hotmail.it; Tel.: +39-071-220-6226
† Presented at the XV National and III International Congress of the Italian Society of Oral Pathology and Medicine (SIPMO), Bari, Italy, 17–19 October 2019.

Published: 11 December 2019

Odontogenic cysts (OC) are one of the most frequent lesions affecting the jaws. These lesions are characterized by a pathologic cavity, either completely or partially covered by an epithelial tissue of odontogenic origin. OCs share similar features; therefore, the differential diagnosis requires a combination of clinical, radiological, and histological findings [1]. This study aims to perform an epidemiologic analysis of OCs treated from 1990 to 2019 at the "Ospedali Riuniti" General Hospital, Ancona, Italy, according to 4th Edition of WHO Classification of Head and Neck Tumours.

The present study considered all the patients who underwent surgery for jaw cysts from January 1990 to August 2019. Data were retrieved and catalogued from clinical records and from the archive of the Institute of Pathology, Marche Polytechnic University, Italy. Because of the 30-year period considered, histological slides of OCs were re-evaluated to confirm the diagnosis, according to the current WHO criteria [2]. From each case, they were extrapolated the following information: age, sex, diagnosis, site distribution, and relapses.

Overall, 1942 patients were treated for jaw cysts, corresponding to 1862 patients with OC, of which 98 showing multiple OCs at the time of diagnosis, and 80 patients with nonodontogenic cysts (NOC). Furthermore, 50 patients showed at least one OC recurrence during follow-up.

2126 surgical specimens were retrieved, corresponding to 2046 OCs and 80 NOCs. 50 patients developed 69 recurrences, mainly Odontogenic keratocysts (OKC). Mean age of occurrence for primary OC was 46.9 ± 17.1 years, with a higher frequency in males (M:F ratio of 1.79). Regarding localization, posterior mandibular and anterior maxillary regions were the most commonly affected sites (Mandible:Maxilla ratio of 1.42). Mean size of primary OC was 1.9 ± 1.0 cm (Table 1).

Radicular cysts were the most frequently diagnosed, with 815 cases (39.83%), followed by Dentigerous cysts (21.51%), and OKC (13.54%) (Figure 1). All other OCs showed a very low frequency, reaching a total of 83 cases (4.06%) (Figure 2). Noteworthy, in 431 cases the clinicopathological data were insufficient to establish a certain diagnosis (21.07%).

Table 1. Demographic and clinical data of OCs (1990–2019).

Clinical Presentation	N° of Cysts
- Primary OCs	1977
- Recurred OCs	69
Site Distribution	
- Mandible	792
- Maxilla	557
- Not specified	697
Sex (n° of Patients)	
- Males	1194
- Females	668
Age (years)	46.9 ± 17.1
Size (cm)	1.9 ± 1.0

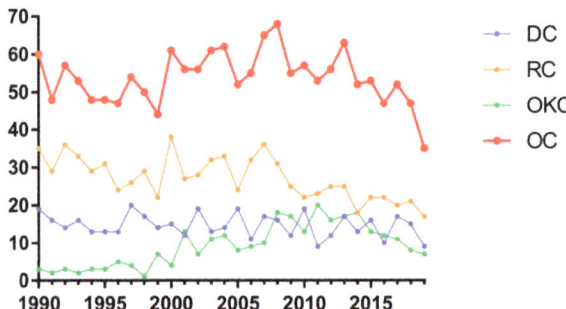

Figure 1. Annual frequency of diagnosed OCs in "Ospedali Riuniti" General Hospital, Ancona, Italy (red line). Radicular cysts (RC, orange line) were the most frequently diagnosed, followed by Dentigerous cysts (DC, blue line) and Odontogenic keratocysts (OKC, green line).

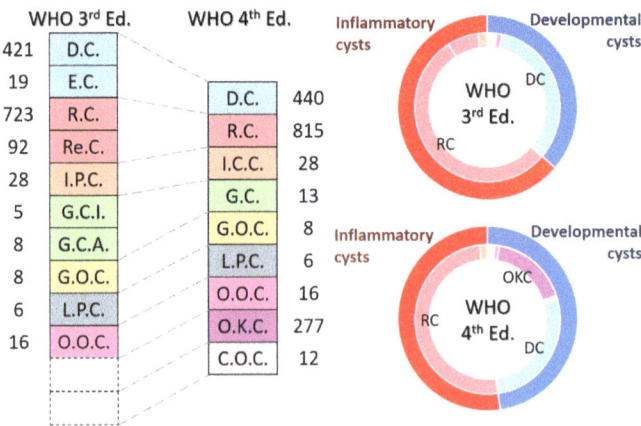

Figure 2. Relative frequency of diagnosed OCs according to 3rd and 4th Edition of WHO Classification, respectively. In 2017 there was a significative simplification of OC classification; the most important changes regard the reintroduction of Odontogenic keratocyst (OKC) and Calcifying Odontogenic cyst (COC). DC = Dentigerous cyst; EC = Eruptive cyst; RC = Radicular cyst; ReC = Residual cyst; IPC = Inflammatory paradental cyst; ICC = Inflammatory collateral cyst; GCI = Gingival cyst of infant; GCA = Gingival cyst of adult; GC = Gingival cyst; GOC = Glandular odontogenic cyst; LPC = Lateral periodontal cyst; OOC = Orthokeratinized odontogenic cyst.

Although limited in its retrospective nature, these findings could be useful to determine the incidence and prevalence of OCs. Prevalence studies related to OCs should be conducted in each tertiary referral center, in order to improve current epidemiological data.

Conflicts of Interest: The authors declare no conflict of interest.

References

1. Lo Muzio, L.; Mascitti, M. Cystic lesions of the jaws: a retrospective clinicopathologic study of 2030 cases. *Oral Surg. Oral Med. Oral Pathol. Oral Radiol.* **2017**, *124*, 128–138, doi:10.1016/j.oooo.2017.04.006.
2. El-Naggar, A.; Chan, J. *WHO Classification of Head and Neck Tumours*, 4th ed.; IARC: Lyon, France, 2017; pp. 232–242.

© 2019 by the authors. Licensee MDPI, Basel, Switzerland. This article is an open access article distributed under the terms and conditions of the Creative Commons Attribution (CC BY) license (http://creativecommons.org/licenses/by/4.0/).

Extended Abstract

13-Gene DNA Methylation Analysis from Oral Brushing: A Non Invasive Diagnostic Tool in the Follow-Up of Patients Surgically Treated for Oral Cancer [†]

Eleonora Morselli [1,*], Roberto Rossi [1], Luca Morandi [2], Achille Tarsitano [3], Andrea Gabusi [1], Linda Sozzi [1], Stesi Kavaja [1] and Davide B Gissi [1]

1. Department of Biomedical and Neuromotor Sciences, Section of Oral Sciences, University of Bologna, 40125 Bologna, Italy; eleonora.morselli@studio.unibo.it (E.L.); roberto.rossi30@studio.unibo.it(R.R.); andrea.gabusi3@studio.unibo.it (A.G.); linda.sozzi2@studio.unibo.it (L.S.); stesi.kavaja@studio.unibo.it (S.K.); davide.gissi@unibo.it (D.B.G.)
2. Functional MR Unit, IRCCS Istituto delle Scienze Neurologiche di Bologna, Department of Biomedical and Neuromotor Sciences, University of Bologna, 40139 Bologna, Italy; luca.morandi2@unibo.it
3. Department of Biomedical and Neuromotor Sciences, Section of Maxillo-facial Surgery at S. Orsola-Malpighi Policlinic, University of Bologna, 40138 Bologna, Italy; achille.tarsitano2@unibo.it
* Correspondence: eleonora.morselli@studio.unibo.it; Tel.: +39-0512088123
† Presented at the XV National and III International Congress of the Italian Society of Oral Pathology and Medicine (SIPMO), Bari, Italy, 17–19 October 2019.

Published: 11 December 2019

1. Introduction

Patients treated for Oral Squamous Cell Carcinoma (OSCC) showed a significant risk to develop a loco-regional relapse during follow-up period. In clinical practice the follow-up strategy for early detection of recurrence or second primary tumors still consists of periodic visual examination and palpation of the oral cavity during the 5-year aftercare period.

Our research group recently developed a non-invasive procedure to identify oral carcinomas at early stage starting from oral brushing, quantitatively measuring the DNA methylation level of a panel of 13 genes [1].

The procedure resulted highly related to the presence of a malignant process (sensitivity 96.6%, specificity 100%). This high accuracy stimulated us to apply our non-invasive diagnostic tool in a cohort of patients previously treated for Oral Squamous Cell Carcinoma. Aim of the study is to evaluate if 13-gene DNA methylation analysis by oral brushing may be a useful procedure to identify patients surgically treated for OSCC at risk of developing a secondary tumor.

2. Materials and Methods

The study population included 49 consecutive surgically treated OSCC. Oral brushing sample collection was performed during patient follow up almost 6 months after OSCC treatment, within the regenerative area after OSCC resection. In all brushing specimens the DNA methylation level of ZAP70, GP1BB, KIF1A, ITGA4, LINC00599, MIR193, MIR296, TERT, LRRTM1, NTM, EPHX3, FLI1 and PARP15 was evaluated by quantitative Bisulfite-Next Generation Sequencing (NGS). Each sample was defined positive or negative in relationship to a specific algorithm and cut-off value [1]. Positive scores in the regenerative area after OSCC resection were analyzed together with other histologic poor prognostic factors for any relationship with appearance of loco-regional relapse.

3. Results

16/49 brushing specimens collected 6 months after surgery from regenerative area after OSCC showed a positive score.

During follow up period 7/49 (14.3%) patients developed a secondary OSCC during follow-up period (mean follow-up: 18.9 months): 6/7 patients showed positive samples from regenerative areas. The presence of a positive score resulted the most powerful variable related to the appearance of locoregional relapse, greater than presence of perineural invasion detected in the surgical OSCC sample.

4. Conclusions

These preliminary results seem to indicate that our novel assay may be proposed as an indicator of disease before appearance of clinical signs and symptoms in surgically treated OSCC patients. Further studies with larger cohort of patients, with adequate follow-up period and brushing sampling collection at different moments are needed to elucidate the prognostic potential of our assay.

Conflicts of Interest: The authors declare no conflict of interest

Reference

1. Morandi, L.; Gissi, D.; Tarsitano, A.; Asioli, S.; Gabusi, A.; Marchetti, C.; Montebugnoli, L.; Foschini, M.P. CpG location and methylation level are crucial factors for the early detection of oral squamous cell carcinoma in brushing samples using bisulfite sequencing of a 13-gene panel. *Clin. Epigenet.* **2017**, *9*, 85.

© 2019 by the authors. Licensee MDPI, Basel, Switzerland. This article is an open access article distributed under the terms and conditions of the Creative Commons Attribution (CC BY) license (http://creativecommons.org/licenses/by/4.0/).

Extended Abstract
Circulating Biochemical Molecular Markers (DNA and RNA) in Head and Neck Cancer: A Narrative Review [†]

Gian Marco Podda *, Federica Rocchetti, Daniele Pergolini, Gaspare Palaia, Gianluca Tenore and Umberto Romeo

Department of Oral and Maxillofacial Sciences, Sapienza University of Rome, 00161 Rome, Italy; federicarocchetti@gmail.com (F.R.); daniele.pergolini@hotmail.it (D.P.); gaspare.palaia@gmail.com (G.P.); gianlucatenore@gmail.com (G.T.); umberto.romeo@uniroma1.it (U.R.)
* Correspondence: g.m.podda@hotmail.it; Tel.: +39-334-959-3848
† Presented at the XV National and III International Congress of the Italian Society of Oral Pathology and Medicine (SIPMO), Bari, Italy, 17–19 October 2019.

Published: 10 December 2019

Head and neck cancers represent the sixth most common cancer group in humans [1]. Amongst these, head and neck squamous cell carcinoma (HNSCC) is the most frequent. In Europe the 5-year survival rate for HNSCC ranges from 25% to 60%, depending on primary tumor site and stage. The objective of a number of ongoing studies is to explore innovative techniques that allow early and reliable diagnosis and at the same time offer the possibility to monitor the evolution of the pathology.

The aim of this narrative review is to evaluate the role of biochemical molecular markers (liquid biopsy), such as DNA and RNA, in the diagnosis, prognostic outcome and treatment-monitoring in patients affected by HNSCC.

A selection of articles published until September 2019 on Pubmed formed the basis of the narrative review. The following inclusion and exclusion criteria were set: studies on human blood, plasma or serum evaluating DNA and RNA expression in patients with HNSCC compared to healthy controls (a minimum of 20 samples for each group needed); studies not including patients affected by Oral Squamous Cell Carcinoma (OSCC) were excluded. Amongst a total of 133 studies found, 8 met inclusion criteria (Figure 1).

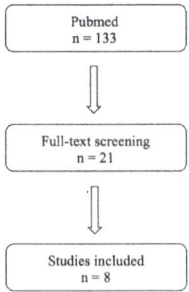

Figure 1. Flow-chart of the selection process for studies included in the narrative review.

Amongst all analysed studies, only one did not find a significant difference in the concentration of the circulating markers between the test group and the control group. Four studies evaluated the concentration of the circulating biochemical markers for the purpose of the diagnosis; one study for the prognostic outcome; one study for both diagnosis and treatment monitoring; two studies evaluated all three outcomes (Table 1).

Table 1. OSCC: Oral Squamous Cell Carcinoma; HSNCC: Head and Neck Squamous Cell Carcinoma; lncRNA: Long Noncoding RNA; cfDNA: Circulating Cell-Free DNA.

First Author and Year	Patients Number	Tumor Stage	Matrix	Cell-Free Nucleic Acid	Detection Method	Assesed for
Yang Li 2006	OSCC: 32 Control: 35	Only Stage I and II	Serum	mRNA	RT-qPCR	Diagnosis
Deepika Shukla 2013	Potentially Malignant Lesions: 90 OSCC: 150 Post-Treatment OSCC: 150 Control: 150	Not reported	Blood and Plasma	cfDNA	NanoDrop spectrophotometer	Diagnosis Treatment monitoring
Ya-Ching Lu 2015	OSCC: 90 Pre-cancer: 16 Control: 53	I-II: 20 III-IV: 60	Plasma	miR-196a miR-196b	RT-qPCR	Diagnosis
Andreas Schröck 2017	HSNCC: 421 Control: 224	T_{is}: 10 N_0: 147 T_1: 87 N_1: 54 T_2: 129 N_2: 140 T_3: 86 N_3: 8 T_4: 80 N_x: 72 NA^d: 29	Plasma	cfDNA (methylation)	qPCR	Diagnosis Prognostic outcome Treatment monitoring
Pooja Singh 2017	OSCC: 20 OSMF: 20 Control: 40	Not reported	Serum	miR-21	RT-qPCR	Prognostic outcome
Li-Han Lin 2018	OSCC: 121 Control: 50	I-II: 40 III-IV: 81	Plasma	cfDNA	spectrophotometer	Diagnosis Prognostic outcome Treatment monitoring
Tingru Shao 2018	OSCC: 80 Control: 70	Not reported	Serum	lncRNA mRNA	RT-qPCR	Diagnosis
Xinyuan Chen 2018	HNSCC: 100 CONTROL: 100	I: 43 II: 31 III: 26 IV: 0	Plasma	lncRNA	NanoDrop spectrophotometer (validation RT-qPCR)	Diagnosis

Although liquid biopsy is an effective tool used by clinicians to approach other cancer types, such as Small Cell Lung Carcinoma (SCLC), its effectiveness with respect to HNSCC has to be investigated to a deeper extent. The biopsy therefore remains the gold standard in the diagnostic process of these cancers. However, liquid biopsy offers an advantage in the monitoring process of treated patients as it is less invasive when compared to Computed Tomography (CT) and Positron Emission Tomography (PET).

In the near future, biochemical molecular markers may lead the way to customized-patient therapy.

Conflicts of Interest: the authors declare no conflict of interest.

Reference

1. Warnakulasuriya, S. Global epidemiology of oral and oropharyngeal cancer. *Oral Oncol.* **2009**, *45*, 309–316.

© 2019 by the authors. Licensee MDPI, Basel, Switzerland. This article is an open access article distributed under the terms and conditions of the Creative Commons Attribution (CC BY) license (http://creativecommons.org/licenses/by/4.0/).

Extended Abstract

Podoplanin Expression and Its Correlation with Perineural Invasion in Oral Squamous Cell Carcinoma [†]

Roberto Rossi [1,*], Achille Tarsitano [2], Sofia Asioli [3], Alice Piastra [1], Andrea Gabusi [1], Luca Morandi [3], Laura Luccarini [1], Laura Felicetti [1] and Davide B. Gissi [1]

1. Department of Biomedical and Neuromotor Sciences, Section of Oral Sciences, University of Bologna, 40159 Bologna, Italy; alice.piastra@gmail.com (A.P.); andrea.gabusi3@unibo.it (A.G.); laura.luccarini2@studio.unibo.it (L.L.); laura.felicetti@unibo.it (L.F.); davide.gissi@unibo.it (D.B.G.)
2. Department of Biomedical and Neuromotor Sciences, Section of Maxillo-Facial Surgery at Policlinico S. Orsola-Malpighi, University of Bologna, 40138 Bologna, Italy; achille.tarsitano2@unibo.it
3. Department of Biomedical and Neuromotor Sciences, Section of Anatomic Pathology at Bellaria Hospital, University of Bologna, 40139 Bologna, Italy; sofia.asioli3@unibo.it (S.A.); luca.morandi2@unibo.it (L.M.)
* Correspondence: roberto.rossi30@studio.unibo.it; Tel.: +39-334-109-9371
† Presented at the XV National and III International Congress of the Italian Society of Oral Pathology and Medicine (SIPMO), Bari, Italy, 17–19 October 2019.

Published: 11 December 2019

1. Introduction

Perineural invasion (PNI) represents cancer propension to spread through neuronal pathways and hinders the ability to establish a surgical local control of Oral Squamous Cell Carcinoma (OSCC). Histologic evidence of PNI is a recognized poor prognostic factor related with an high risk of loco-regional recurrence and/or presence of an occult lymphnode metastasis.

Unfortunately, PNI can be evaluated only in surgical specimens of OSCC and not in preoperative incisional biopsies, rendering timely therapeutic planning impossible.

Identification of a pre-operative reliable molecular marker related with presence of PNI may represent an attractive strategy. In this sense, recent studies showed that podoplanin is overexpressed in aggressive tumors and is associated with cervical lymphatic dissemination and poor survival in OSCC [1].

The aim of this study is to evaluate the relationship between Podoplanin altered expression in pre-operative incisional biopsy and presence of PNI in surgical OSCC sample and their prognostic role.

2. Materials and Methods

The cohort of this study consisted of 85 consecutive OSCC patients (average follow-up: 21.46 months). In all cases a pre-operative incisional biopsy for histological assessment and immunohistochemical analysis of Popoplanin was performed. PNI evidence was histologically investigated in all samples after surgical exeresis of OSCC.

3. Results

During follow-up period, 27/85 (31.7%) OSCC patients developed a second neoplastic event and 15/85 (17.7%) patients died due to malignancy.

32/85 (37.7%) OSCC surgical samples showed presence of PNI in surgical sample. In the population study PNI resulted the only independent clinico-pathological variable significantly related with appearance of a loco-regional recurrence and with disease-specific survival.

Podoplanin overexpression was found in 44/85 (51.7%) in pre-operative incisional biopsies.

Multiple Logistic Regression showed that Podoplanin overexpression in incisional biopsy samples is the only pre-operative variable significantly related to presence of PNI, indeed an altered expression of podoplanin was found in 30/32 PNI positive OSCCs with respect to 14/53 PNI negative OSCCs.

4. Conclusions

The present study confirmed and highlighted the prognostic role of PNI in OSCC patients and for the first time demonstrated that a positive expression of Podoplanin in pre-operative biopsy is significantly related with PNI status. Indeed, podoplanin showed high sensitivity and good performances as negative predictive marker for PNI presence and may be helpful for an accurate preoperative risk stratification of patients with OSCC.

Conflicts of Interest: The authors declare no conflict of interest.

Reference

1. Foschini, M.P.; Leonardi, E.; Eusebi, L.H.; Farnedi, A.; Poli, T.; Tarsitano, A.; Cocchi, R.; Marchetti, C.; Gentile, L.; Sesenna, E.; et al. Podoplanin and E-cadherin expression in preoperative incisional biopsies of oral squamous cell carcinoma is related to lymph node metastases. *Int. J. Surg. Pathol.* **2013**, *21*, 133–141.

© 2019 by the authors. Licensee MDPI, Basel, Switzerland. This article is an open access article distributed under the terms and conditions of the Creative Commons Attribution (CC BY) license (http://creativecommons.org/licenses/by/4.0/).

Extended Abstract

Osteoradionecrosis Rate in Patients Undergoing Radiotherapy for Head and Neck Cancer Treatment: A Six Months Follow-Up of a Perspective Clinical Study [†]

Cosimo Rupe [1,*], Francesco Miccichè [2], Gaetano Paludetti [3], Patrizia Gallenzi [1] and Carlo Lajolo [1]

1. Head and Neck Department, "Fondazione Policlinico Universitario A. Gemelli–IRCCS", School of Dentistry, Università Cattolica del Sacro Cuore, 00168 Rome, Italy; patrizia.gallenzi@unicatt.it (P.G.); carlo.lajolo@unicatt.it (C.L.)
2. Department of Radiation Oncology, "Fondazione Policlinico A. Gemelli–IRCCS", Institute of Radiology, Università Cattolica del Sacro Cuore, 00168 Rome, Italy; francesco.micciche@unicatt.it
3. Head and Neck Department, "Fondazione Policlinico Universitario A. Gemelli–IRCCS", Institute of Otolaryngology, Università Cattolica del Sacro Cuore, 00168 Rome, Italy; gaetano.paludetti@unicatt.it
* Correspondence: cosimorupe@gmail.com; Tel.: +39-392-938-1949
† Presented at the XV National and III International Congress of the Italian Society of Oral Pathology and Medicine (SIPMO), Bari, Italy, 17–19 October 2019.

Published: 10 December 2019

High-dose radiotherapy (RT) for head and neck cancer has significant adverse effects on maxillofacial tissues, among which osteoradionecrosis (ORN) is the most severe and potentially life-threatening. Although tooth extractions seem to be the main risk factor, few perspective studies evaluated protocols to minimize the ORN risk due to extractions [1]. The aim of this study is to evaluate incidence and risk factors of ORN in a cohort of patients receiving tooth extractions before RT and evaluate an algorithm about extraction decision.

One-hundred and twenty-eight patients were recruited in this study: impacted third molars with radiographic sign of pericoronitis, teeth with periapical lesions, unrestorable teeth, periodontally compromised teeth (pocket probing depth > 5 mm, clinical attachment loss > 8 mm, grade 2 tooth mobility, II grade furcation involvement) were extracted under antibiotic prophylaxis.

A 15-days interval between the last tooth extraction and the beginning of RT was recommended. Patients were visited at 15 days, 1, 3 and 6 months after the beginning of RT. Data of patients with a minimum of 6 months follow-up are presented in this report.

ORN was defined as irradiated exposed necrotic bone, without healing for 3 months, in absence of cancer recurrence, and staged according to Notani et al. [2]. RT treatment plan were reviewed, and each post-extractive socket was contoured with the software Eclipse Treatment Planning System (Varian Medical System). The dose received by each socket was finally calculated. The protocol was approved by the Ethic Committee of Catholic University–Fondazione Policlinico Gemelli (Prot. OHHN-1, ID-2132).

Out of 128 patients, 64 had a 6 months follow-up, 41 of whom received at least one tooth extraction, for a total of 183 teeth. Twenty-one days was the mean time interval before the beginning of RT. Three sites in 3 different patients (7.3% of patients) developed ORN, all in the posterior mandible and two of them also received chemotherapy.

Tooth extraction has been recognized as a risk factor for the development of ORN, especially in the posterior mandible. In the preliminary results of our study, 3 ORN happened and the major risk factor was the anatomical site. These results are consistent with the current literature, suggesting that the proposed protocol can be a valid support in the decision-making process. Since ORN can be a late onset complication of RT, a longer follow-up must be performed to draw definitive conclusions.

Conflicts of Interest: The authors declare no conflict of interest.

References

1. Nabil, S.; Samman, N. Risk factors for osteoradionecrosis after head and neck radiation: A systematic review. *Oral Surg. Oral Med. Oral Pathol. Oral Radiol.* **2012**, *113*, 54–69.
2. Notani, K.I.; Yamazaki, Y.; Kitada, H.; Sakakibara, N.; Fukuda, H.; Omori, K.; Nakamura, M. Management of mandibular osteoradionecrosis corresponding to the severity of osteoradionecrosis and the method of radiotherapy. *Head Neck* **2003**, *25*, 181–186.

© 2019 by the authors. Licensee MDPI, Basel, Switzerland. This article is an open access article distributed under the terms and conditions of the Creative Commons Attribution (CC BY) license (http://creativecommons.org/licenses/by/4.0/).

Extended Abstract

Teeth Extractions in Subjects Undergoing Radiotherapy for Head and Neck Cancers: A Systematic Review on the Clinical Protocols for Preventing Osteoradionecrosis (ORN). Extractions before Radiotherapy (Part 1) †

Cosimo Rupe [1,*], Gioele Gioco [1], Giuseppe Troiano [2], Michele Giuliani [2], Maria Contaldo [3] and Carlo Lajolo [1]

1. Head and Neck Department, "Fondazione Policlinico Universitario A. Gemelli–IRCCS", School of Dentistry, Università Cattolica del Sacro Cuore, 00168 Rome, Italy; gioele.gioco@hotmail.it (G.G.); carlo.lajolo@unicatt.it (C.L.)
2. Department of Clinical and Experimental Medicine, University of Foggia, 71121 Foggia, Italy; giuseppe.troiano@unifg.it (G.T.); michele.giuliani@unifg.it (M.G.)
3. Department of Medical-Surgical and Odontostomatological Specialties, University of Campania "Luigi Vanvitelli", 80138 Naples, Italy; maria.contaldo@gmail.com
* Correspondence: cosimorupe@gmail.com; Tel.: +39-3929381949
† Presented at the XV National and III International Congress of the Italian Society of Oral Pathology and Medicine (SIPMO), Bari, Italy, 17–19 October 2019.

Published: 10 December 2019

Osteoradionecrosis (ORN) of the jaws is the most severe side effect, in some case life-threatening, of radiotherapy for head and neck cancer [1]. Tooth extractions before radiotherapy seems to lead to a reduction of ORN onset, nevertheless the real ORN rate and its risk factors are still unclear [2]. The aim of this systematic review is to determine the rate of ORN due to tooth extraction before the beginning of RT in patients affected by head and neck cancer and to identify any possible risk factor for ORN onset.

PRISMA protocol was used to evaluate and present the results. PubMed, Scopus, and Web of Science were used as search engines: only English full-length papers of clinical trial and observational studies both prospective and retrospective, published in peer-reviewed journals, were investigated. Inclusion criteria: minimum sample size 10 patients who underwent extraction before radiotherapy in head and neck district; minimum 6 months of follow-up after RT; ORN diagnosis criteria clearly defined in the text; specified if ORN developed at extraction site or not. Cumulative meta-analysis was performed calculating the pooled proportion (PP) of ORN occurrence rate. Meta-analysis was performed at random effects model with the Der-Simonian Liard method. All the statistical analyses were performed with the software Open Meta-Analyst version 10. PROSPERO registration code is CRD42018079986.

Among 2020 records screened, only 8 were included in this review (Figure 1). Sixteen of 494 patients who underwent tooth extraction before radiotherapy developed ORN at extraction site, with an ORN incidence of 2.2% (95% Confidence of Interval = 0.8—3.9, $p < 0.204$, $I^2 = 2695\%$) (Figure 2). All cases of ORN due to pre-RT dental extraction were reported in mandible.

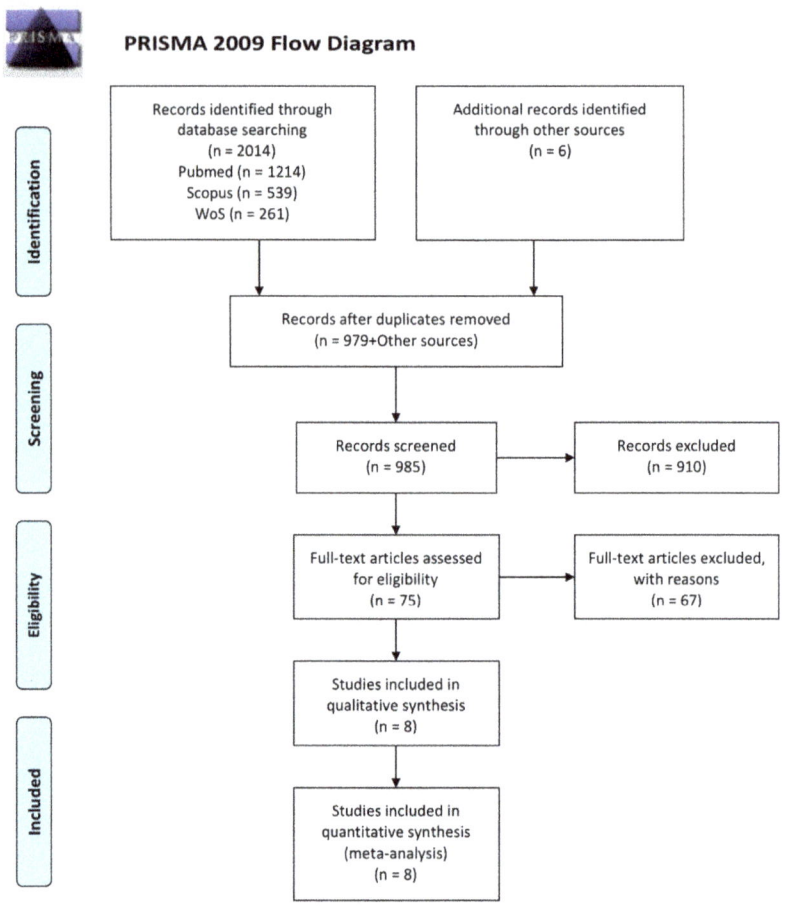

Figure 1. PRISMA flow-chart of the inclusion process.

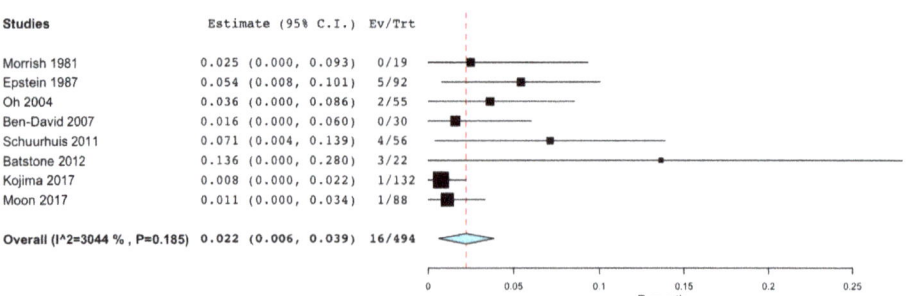

Figure 2. Pooled data and box plot stratified for studies.

The considerable amount of missing data didn't allow us to identify other possible risk factor for ORN onset (i.e., surgical protocol, time interval between the extraction and the beginning of radiotherapy, dose). Major efforts should be done to perform sounder methodological clinical investigations. It is general recommendation that patients should undergo a dental examination and oral foci should be removed before the beginning of radiotherapy for head and neck cancer, in order to reduce the risk of occurrence of undesirable effects. Although tooth extraction prior to radiotherapy leads to a reduction of ORN incidence, this systematic review shows that ORN risk is still high and special attention should be posed to extraction site.

Conflicts of Interest: The authors declare no conflict of interest.

References

1. Marx, R.; Johnson, R. Studies in the radiobiology of osteoradionecrosis and their clinical significance. *Oral Surg. Oral Med. Oral Pathol.* **1987**, *64*, 379–390.
2. Chronopoulos, A.; Zarra, T.; Ehrenfeld, M.; Otto, S. Osteoradionecrosis of the jaws: Definition, epidemiology, staging and clinical and radiological findings. A concise review. *Int. Dent. J.* **2018**, *68*, 22–30.

© 2019 by the authors. Licensee MDPI, Basel, Switzerland. This article is an open access article distributed under the terms and conditions of the Creative Commons Attribution (CC BY) license (http://creativecommons.org/licenses/by/4.0/).

Extended Abstract

Prognostic Role of DNA Methylation Analysis from Oral Brushing in Oral Squamous Cell Carcinoma [†]

Valentina Russo [1,*], Jacopo Lenzi [2], Roberto Rossi [1], Luca Morandi [3], Achille Tarsitano [4], Andre Gabusi [1], Chiara Amadasi [1], Dora Servidio [1] and Davide B. Gissi [1]

[1] Section of Oral Sciences, Department of Biomedical and Neuromotor Sciences, University of Bologna, 40125, Bologna, Italy; roberto.rossi30studio.unibo.it (R.R.); andrea.gabusi3@unibo.it (A.G.); chiama90@live.it (C.A.); dora.servidio@gmail.com (D.S.); davide.gissi@unibo.it (D.B.G.)
[2] Section of Hygiene, Public Health and Medical Statistics, Department of Biomedical and Neuromotor Sciences, University of Bologna, 40126 Bologna, Italy; jacopo.lenzi2@unibo.it
[3] Functional MR Unit, Department of Biomedical and Neuromotor Sciences, IRCCS Istituto delle Scienze Neurologiche di Bologna, University of Bologna, 40139 Bologna, Italy; luca.morandi2@unibo.it
[4] Section of Maxillo-Facial Surgery at S. Orsola-Malpighi Policlinic, Department of Biomedical and Neuromotor SciencesUniversity of Bologna, 40138, Bologna, Italy; achille.tarsitano2@unibo.it
* Correspondence: valentina.russo15@studio.unibo.it; Tel: +39-051-208-8123
[†] Presented at the XV National and III International Congress of the Italian Society of Oral Pathology and Medicine (SIPMO), Bari, Italy, 17–19 October 2019.

Published: 11 December 2019

1. Introduction

The aim of this study was to investigate the prognostic role of methylation profile of 13 genes starting from oral brushing collected in a group of patients affected by oral squamous cell carcinoma (OSCC). To this purpose, we analyzed the relationship between the methylation profile of each gene and the appearance of a loco-regional relapse in a study sample of OSCC patients. We also calculated a unique prognostic score with the aim to predict survival of OSCC patients.

2. Materials and Methods

The study sample included 37 consecutive OSCC patients with a median follow-up period of 22 months (range 1–59).

Brushing cell collection was performed prior to any cancer treatment in the tumor mass. Two hundred forty-five CpG islands from a set of 13 previously described methylated genes in OSCC (ZAP70, KIF1A, LRRTM1, PARP15, FLI1, NTM, LINC0059, EPHX3, ITGA4, MIR193, GP1BB, MIR296, TERT) [1] were investigated by bisulfite-Target Next Generation Sequencing (NGS) using MiSEQ platform (Illumina, San Diego, CA, USA).

Univariable Cox proportional hazards models were used to analyze the association between each of the CpG sites and survival. Then, a Cox proportional hazards lasso model was used to select the prognostic markers of the candidate CpG sites. Lastly, a cross-validated prognostic score was computed for each patient, based on individual values of methylation and non-zero regression coefficients. Kaplan-Meier estimates were calculated to compare the survivor functions of the two groups (High-risk of relapse group and low-risk of relapse group). Statistical significance of the log-rank test was set at 0.05.

3. Results

Nine out of 37 (24%) OSCC patients developed a secondary neoplastic manifestation during the follow-up period. Cox proportional hazards lasso model selected 5 CpG sites significantly related with appearance of a second neoplastic event (ZAP70-position1, FLI1-position3, FLI1-position4,

ITGA4-position4 and MIR193-position3). A prognostic score for each patient was calculated, and OSCC patients were divided based on risk of relapse (high and low risk): 8/18 high-risk group patients developed a local relapse with respect to 1/19 low-risk group patients, this difference being statistically significant ($p < 0.001$).

4. Conclusions

Our study showed that a prognostic score based on DNA methylation analysis might be a useful indicator in surgical decision making, even if a larger cohort of patients is necessary to confirm these preliminary data.

Conflicts of Interest: The authors declare no conflict of interest.

References

1. Morandi, L.; Gissi, D.; Tarsitano, A.; Asioli, S.; Gabusi, A.; Marchetti, C.; Montebugnoli, L.; Foschini, M.P. CpG location and methylation level are crucial factors for the early detection of oral squamous cell carcinoma in brushing samples using bisulfite sequencing of a 13-gene panel.Clin Epigenetics. *Clin. Epigenet.* **2017**, *9*, 85

© 2019 by the authors. Licensee MDPI, Basel, Switzerland. This article is an open access article distributed under the terms and conditions of the Creative Commons Attribution (CC BY) license (http://creativecommons.org/licenses/by/4.0/).

Extended Abstract

TIMELESS in Head and Neck Squamous Cell Carcinoma: A Systematic Review [†]

Piermichele Saracino [1],*, Claudia Arena [1], Marco Mascitti [2], Andrea Santarelli [2], Vera Panzarella [3] and Lucio Lo Russo [1]

1. Department of Clinical and Experimental Medicine, University of Foggia, 71122 Foggia, Italy; claudia.arena@unifg.it (C.A.); lucio.lorusso@unifg.it (L.L.R.)
2. Department of Clinical Specialistic and Dental Sciences, Marche Polytechnic, 60131 Ancona, Italy; MarcoMascitti86@hotmail.it (M.M.); andrea.santarelli@staff.univpm.it (A.S.)
3. Department of Surgical, Oncological and Oral Sciences, University of Palermo, 90127 Palermo, Italy; vera.panzarella@unipa.it
* Correspondence: piermichele.saracino@unifg.it; Tel.: +393332016671
† Presented at the XV National and III International Congress of the Italian Society of Oral Pathology and Medicine (SIPMO), Bari, Italy, 17–19 October 2019.

Published: 11 December 2019

TIMELESS is one of the main circadian genes. Different roles are described, such as replication fork stability, cell survival after DNA damage or replication stress by promoting DNA repair. TIMELESS deficiency increases genomic instability and its reduction increases Rad51 and Rad52 foci formation during S phase [1]. TIMELESS and PARP1 operate in a complex to mediate DNA repair. It is also showed that alteration in circadian rhythm is associated with cancer development and tumor progression, such as chronic myeloid leukemia, hepatocellular carcinoma, and breast cancer [2]. We wanted to summarize the role of TIMELESS in head and neck squamous cell carcinoma. To do so, we performed a literature review using these keywords: TIMELESS [All Fields] AND ("neoplasms" [MeSH Terms] OR "neoplasms" [All Fields] OR "cancer" [All Fields]). We found 54 studies. At the end of the selection process, 4 studies were considered suitable for the analysis of TIMELESS in HNSCC. In Ao et al. TIMELESS resulted downregulated in a HNSCC cell line. In Bjarnason's study [3] through TS Enzyme Activity confirmed its downregulation in biopsy model. Hsu in 2011 [4] used real-time quantitative RTN-PCR on 40 tissue samples with HNSCC to clarify whether the expression levels of circadian clock genes were altered in tumor tissues. Data demonstrated that the expression levels of PER1, PER2, PER3, CRY2, and BMAL were significantly downregulated in HNSCC ($p < 0.01$) (Figure 1), showing its dysregulation in cancer. TIMELESS seems to play an important role in cancer and further investigations are needed to investigate its role as prognostic marker or pharmaceutical target.

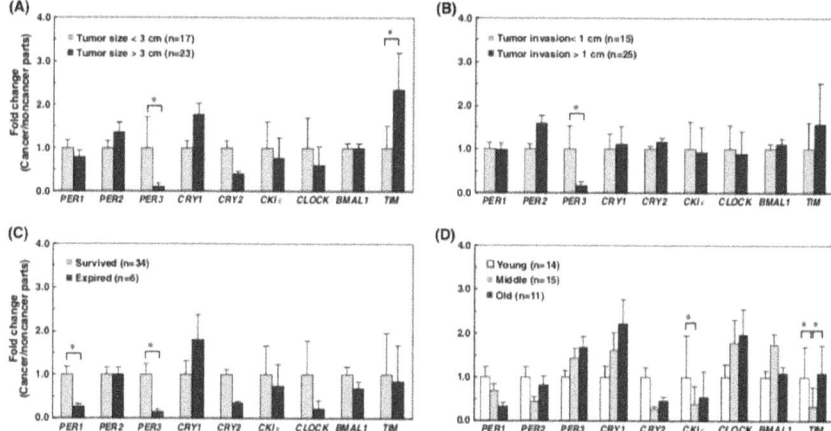

Figure 1. Tumor conditions and circadian clock gene expression in HNSCC patients. Tumor size (**A**), tumor invasion (**B**), survival (**C**) and age (**D**) of the 40 HNSCC patients were correlated to the expression of the nine circadian clock genes. The y axis represent the relative mRNA expression level. The relative expression in cancerous tissues is calculated by ΔΔCt. The expression in tumor size < 3 cm (**A**), tumor invasion < 1 cm (**B**) and survival (**C**) is designated 1, and the relative expression in tumor size > 3 cm (**A**), tumor invasion >1 cm (**B**), and expire (**C**) is calibrated to obtain the folds changed, respectively Statistical significance * $p < 0.05$ evaluated with t test.

Conflicts of Interest: The authors declare no conflict of interest.

References

1. Ao, Y.; Zhao, Q.; Yang, K.; Zheng, G.; Lv, X.; Su, X. A role for the clock period circadian regulator 2 gene in regulating the clock gene network in human oral squamous cell carcinoma cells. *Oncol. Lett.* **2018**, *15*, 4185–4192.
2. Young, L.M.; Marzio, A.; Perez-Duran, P.; Reid, D.A.; Meredith, D.N.; Roberti, D.; Star, A.; Rothenberg, E.; Ueberheide, B.; Pagano, M. TIMELESS forms a complex with PARP1 distinct from its complex with TIPIN and plays a role in the DNA damage response. *Cell Rep.* **2015**, *13*, 451–459.
3. Bjarnason, G.A.; Jordan, R.C.; Wood, P.A.; Li, Q.; Lincoln, D.W.; Sothern, R.B.; Hrushesky, W.J.M.; Ben-David, Y. Circadian expression of clock genes in human oral mucosa and skin: association with specific cell-cycle phases. *Am. J. Pathol.* **2001**, *158*, 1793–1801.
4. Hsu, C.M.; Lin, S.F.; Lu, C.T.; Lin, P.M.; Yang, M.Y. Altered expression of circadian clock genes in head and neck squamous cell carcinoma. *Tumor Biol.* **2012**, *33*, 149–155.

© 2019 by the authors. Licensee MDPI, Basel, Switzerland. This article is an open access article distributed under the terms and conditions of the Creative Commons Attribution (CC BY) license (http://creativecommons.org/licenses/by/4.0/).

Extended Abstract

Salivary ¹H-NMR Metabolomics in Primary Sjögren Syndrome. Preliminary Results of a Pilot Case-Control Study [†]

Giacomo Setti [1,2,*], Gilda Sandri [3], Elisabetta Tarentini [4], Lucia Panari [1], Adele Mucci [5], Valeria Righi [5,6], Marco Meleti [7], Cristina Magnoni [4], Ugo Consolo [1] and Pierantonio Bellini [1]

1. Dentistry and Oral-Maxillofacial Surgery Unit, University of Modena and Reggio Emilia, 41124 Modena, Italy; lucia.panari@hotmail.it (L.P.); ugo.consolo@unimore.it (U.C.); pierantonio.bellini@unimore.it (P.B.)
2. Molecular Medicine Ph.D. School, University of Parma, 43121 Parma, Italy
3. Rheumatology Unit, University of Modena and Reggio Emilia, 41124 Modena, Italy; gilda.sandri@unimore.it
4. Dermatology Unit, Laboratory of Cutaneous Biology, Chi.Mo.Mo Department, University of Modena and Reggio Emilia, 41124 Modena, Italy; elisabetta.tarentini@gmail.com (E.T.); magnoni.cristina@gmail.com (C.M.)
5. Department of Chemical and Geological Sciences, University of Modena and Reggio Emilia, 41125 Modena, Italy; adele.mucci@unimore.it (A.M.); valeria.righi2@unibo.it (V.R.)
6. Department for Life Quality Studies, University of Bologna, 47921 Rimini, Italy; valeria.righi2@unibo.it
7. Oral Medicine and Oral Surgery Laser Unit, University Center of Dentistry, University of Parma, 43121 Parma, Italy; marco.meleti@unipr.it
* Correspondence: giacomo.setti@unipr.it; Tel.: +39-340-666-3903
† Presented at the XV National and III International Congress of the Italian Society of Oral Pathology and Medicine (SIPMO), Bari, Italy, 17–19 October 2019.

Published: 11 December 2019

Primary Sjögren Syndrome (pSS) is a multisystem autoimmune disease which mainly involves exocrine glands, such as salivary and lacrimal. Pathogenesis is not completely understood even if is distinguished by lymphocyte glands tissue infiltration, which leads to anatomical modification and hypofunction. Age at diagnosis is typically between 3rd and 5th decade. European incidence *per* year is 3 to 11 cases on 100.000 subjects, with a female-male ratio of 10:1 and a 0.01% e 0.72% prevalence.

Diagnostic criteria are resumed in the 2016 ACR/EULAR Consensus of Classification Criteria for pSS, and are based on the sum of weighted scores applied to 5 items: anti-SSA/Ro antibody positivity, focus score of ≥1 foci/4 mm², abnormal ocular staining, a Schirmer's test result of ≤5 mm/5 min and an unstimulated salivary flow rate (SFR) of ≤0.1 mL/minute. The prognosis of pSS is favorable with a patient's life expectancy comparable with general population. Quality of life is reduced by the diverse manifestations of the disease. Cardiovascular disease, infections, solid tumors, and lymphoma are the main causes of death [1].

The aim of this pilot study is to compare the salivary metabolome of pSS and healthy controls (HC). Cases were selected from a cohort of pSS patients; age and sex matched HC were recruited from a cohort of volunteers. Strict inclusion and exclusion criteria were applied, in order to select the most homogeneous study population. To all recruited patients was asked to not eat, drink, smoke and use oral hygiene products one hour before saliva collecting. Whole unstimulated saliva was collected into a sterile Eppendorf. SFR was contextually evaluated. All samples were immediately frozen into liquid nitrogen and stored into −80 °C refrigerator until analysis.

7 pSS and 6 HC female patients were recruited. All samples were centrifuged at 15,000 rpm/10 min, 4 °C. 100 μL of buffer were added to 500 μL of supernatant; the solution was inserted in the

NMR tube of a Bruker Avance III HD 600 MHz spectrometer for the ¹H-NMR analysis. Mono-dimensional and bi-dimensional measurements were performed using different pulse sequences such as Carr-Purcell-Meiboom-Gill (cpmg) and different correlation sequence such as COrrelation SpectroscopY (COSY) and Heteronuclear Single Quantum Coherence (HSQC) (Figure 1). Software aided data analysis were performed.

33 metabolites were detected. An high metabolite variability was observed. Normalized spectral matrix multivariate statistical analysis (PCA and PLS-DA) returned interesting results, describing significant differences of metabolites expressions between groups (Figure 1).

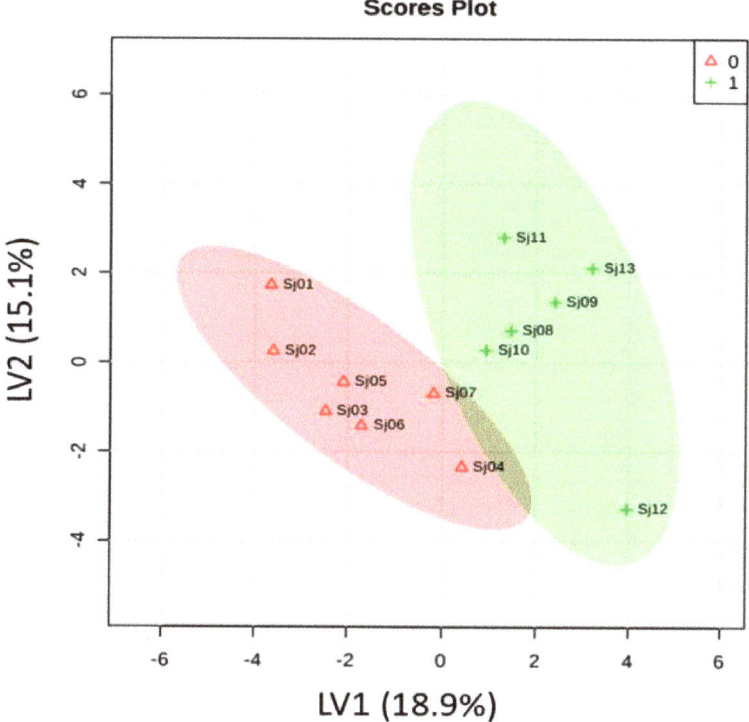

Figure 1. Partial Least Squares Discriminant Analysis (PLS-DA). Chemo-metric differences between groups emerged (SS vs. HC); algorithm supervised analysis showed separation of variables' clusters (metabolites).

Conflicts of Interest: The authors declare no conflict of interest.

Reference

1. Shiboski, C.H.; Shiboski, S.C.; Seror, R.; Criswell, L.A.; Labetoulle, M.; Lietman, T.M.; Rasmussen, A.; Scofield, H.; Vitali, C.; Bowman, S.J.; et al. 2016 American College of Rheumatology/European League Against Rheumatism classification criteria for primary Sjögren's syndrome. *Ann. Rheum. Dis.* **2017**, *76*, 9–16, doi:10.1136/annrheumdis-2016-210571.

© 2019 by the authors. Licensee MDPI, Basel, Switzerland. This article is an open access article distributed under the terms and conditions of the Creative Commons Attribution (CC BY) license (http://creativecommons.org/licenses/by/4.0/).

Extended Abstract

Gum Hypertrophy in Patients in Fixed Orthodontic Therapy Treated with Topical Probiotic Lactobacillus Reuteri: A Pilot Study [†]

Antonia Sinesi [1],*, Giovanna Mosaico [2], Martina Cont [3], Savino Cefola [4], Giovanni Mautarelli [5] and Cinzia Casu [6]

1. RDH, Freelancer, 76012 Canosa di Puglia, Italy
2. RDH, Freelancer, 76012 Brindisi, Italy; gimosaico@tiscali.it
3. RDH, Freelancer, 38014 Trento, Italy; contmartina@gmail.com
4. DDS, Private Dental Practice, 76012 Barletta, Italy; info@drsavinocefola.it
5. DDS, Private Dental Practice, 76112 Brindisi, Italy; studiomautarelli@tiscali.it
6. DDS, Private Dental Practice, 09126 Cagliari, Italy; ginzia.85@hotmail.it
* Correspondence: antonia.sinesi@gmail.com
† Presented at the XV National and III International Congress of the Italian Society of Oral Pathology and Medicine (SIPMO), Bari, Italy, 17–19 October 2019.

Published: 12 December 2019

1. Introduction

Patients undergoing in fixed orthodontic treatment could have higher risk in periodontal/gum disease development, such as gingival hypertrophies. Different types of therapies have been proposed for the treatment of this condition, such as traditional surgery, laser therapy and the use of chlorhexidine mouthwashes [1–3]. The aim of this pilot study is to evaluate the use of a topical probiotic with Lactobacillus Reuteri in the treatment of patients with gingival hypertrophy in orthodontic therapy.

2. Materials and Methods

For this study, 14 patients (10 females and 4 males) with gingival hypertrophy of the incisal group in fixed orthodontic therapy, were considered (Figure 1). Topical probiotics with Lactobacillus reuteri DSM 17,938 and ATCC PTA 5289 were used to treat gingival hyperplasia. At the first session periodontal clinical parameters were recorderd (T0) and after debridement, topical application were made. The probiotic gel is made by powdered tablets in a 1% Gel Carbopol; The gel was injected into the pockets until the spill, for 3–5 min the patient did not rinse the mouth and was advised not to eat and drink for about an hour. Systemic probiotics of Lactobacillus Reuteri taken at home 2 times a day were prescribed. After 1-month periodontal clinical parameters were re-evaluated (T1).

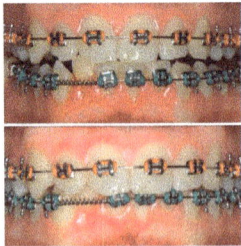

Figure 1. 19 years old male patient before and after treatment.

3. Results

The mean age of the patient was 14.8 years old. All patients showed a complete regression of gingival hypertrophies (Figure 2) at one month of follow up (100%). The mean value of plaque index (PI) before therapy was 69.71%, while at one month it was 18.57%.

The average value of the bleeding index (BoP) was 37.85% while at one month after treatment it was 2.35%.

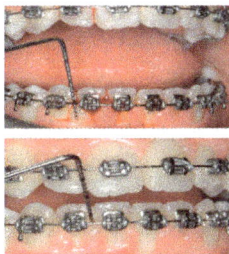

Figure 2. 14 years old female patient before and after the treatment.

4. Discussion and Conclusions

All gingival lesions regressed after treatment, while PI was reduced by more than 3.5 times. The average BoP values are reduced so as to be negligible. A previous study has tested the use of two mouthwashes for the treatment of this condition, but the authors concluded that "none of these principals, although the gingival condition improved, could reduce gingival enlargement to the clinically acceptable level of health [1]". Other works suggest a surgical removal, with or without laser of the excess gum tissue [2,3]. No other works in the literature has used topical probiotics for the treatment of gingival hypertrophies in orthodontic patients, which from our observation proved to be effective and safe because they had no side effects. Studies with a larger sample and with greater follow-up are necessary to confirm these preliminary results.

Reference

1. Farhadian, N.; Bidgoli, M.; Jafari, F.; Mahmoudzadeh, M.; Yaghobi, M.; Miresmaeili, A. Comparison of Electric Toothbrush, Persica and Chlorhexidine Mouthwashes on Reduction of Gingival Enlargement in Orthodontic Patients: A Randomised Clinical Trial. *Oral Health Prev. Dent.* **2015**, *13*, 301–307.
2. Jadhav, T.; Bhat, K.M.; Bhat, G.S.; Varghese, J.M. Chronic inflammatory gingival enlargement associated with orthodontic therapy case report. *J. Dent. Hyg.* **2013**, *87*, 19–23.
3. Gama, S.K.; De Araújo, T.M.; Pozza, D.H.; Pinheiro, A.L. Use of the CO_2 laser on orthodontic patients suffering from gingival hyperplasia. *Photomed. Laser Surg.* **2007**, *25*, 214–219.

 © 2019 by the authors. Licensee MDPI, Basel, Switzerland. This article is an open access article distributed under the terms and conditions of the Creative Commons Attribution (CC BY) license (http://creativecommons.org/licenses/by/4.0/).

Extended Abstract

Adenoid Cystic Carcinoma of Salivary Gland: An Immunohistochemical Study [†]

Lucrezia Togni [1,*], Marco Mascitti [1], Corrado Rubini [2], Rodolfo Mauceri [3], Lorenzo Lo Muzio [4] and Andrea Santarelli [1]

1. Department of Clinical Specialistic and Dental Sciences, Marche Polytechnic University, 60126 Ancona, Italy; marcomascitti86@hotmail.it (M.M.); andrea.santarelli@staff.univpm.it (A.S.)
2. Department of Biomedical Sciences and Public Health, Marche Polytechnic University, 60126 Ancona, Italy; c.rubini@univpm.it
3. Department of Surgical, Oncological and Oral Sciences (DICHIRONS), University of Palermo, 90127 Palermo, Italy; rodolfo.mauceri@unipa.it
4. Department of Clinical and Experimental Medicine, University of Foggia, 71122 Foggia, Italy; lorenzo.lomuzio@unifg.it
* Correspondence: togni.lucrezia@gmail.com; Tel.: +39-071-2206226
† Presented at the XV National and III International Congress of the Italian Society of Oral Pathology and Medicine (SIPMO), Bari, Italy, 17–19 October 2019.

Published: 11 December 2019

Adenoid cystic carcinoma (ACC) is a basaloid tumour consisting of epithelial and myoepithelial cells in variable morphologic configurations, including tubular, cribriform and solid patterns. ACC accounts for 1% of head and neck cancers and 10% of salivary gland neoplasms. It is characterized by invasive growth, perineural infiltration, early local recurrence and late onset distant metastasis. Radical surgical excision, with or without postoperative radiotherapy, is the treatment of choice. 5-year survival rate is approximately 35% and local recurrences occur despite combined treatment [1]. Thus, novel therapies are urgently needed to supplement the treatment currently available. The study of the molecular mechanisms involved in the neoplasm development is necessary to identify new prognostic markers.

This study investigated the immunohistochemical expression of a panel of molecular markers (C-Kit, Ki-67, p53, and E-Cadherin) to evaluate the correlation with clinicopathological data and prognostic factors, to lay the foundation for further studying the pathogenesis and treatment of the ACC.

The study included surgical resection specimens obtained from 17 ACC. Data were retrieved and cataloged from clinical records and from the archive of the Institute of Pathology, Marche Polytechnic University. 4-μm serial sections were incubated with the polyclonal rabbit anti-human CD117 antibody (Agilent-Dako), diluted 1:50; with the monoclonal mouse anti-human Ki-67 antigen antibody (Agilent-Dako), diluted 1:50; with the FLEX monoclonal mouse anti-human p53 protein antibody (Dako Omnis); and with the FLEX monoclonal mouse anti-human E-Cadherin antibody (Dako Omnis), in a humidified chamber at room temperature for 1 h. To evaluate the extension of markers expression, the mean percentage of positive cells was determined from the analysis of 1000 cells at ×40 magnification. Staining intensity of CD117 and p53, were scored as "-" (negative staining); "+" (weak staining); "++" (moderate staining); "+++" (intense staining).

No significant associations between the evaluated markers and clinicopathological data were demonstrated. Closely related but not significant correlations were found analyzing the p53 immunohistochemical expression. A lower expression was observed in cribriform compared to solid growth pattern (10.4% vs 38.7%) and in I-II respect to III-IV stage-disease (14.9% vs 45.2%).

The results not reveal a prognostic role for these molecular markers. However, p53 expression could be correlated to worst prognosis. The lack of relationships could be attributed the low records available, due to the rarity of this malignancy. So, further studies on larger series should be conducted. Literature data are extremely heterogeneous and although important advances have been made in exploring the biological bases of ACC, many critical factors remain elusive. Therefore, explore the ACC prognostic indicators is necessary for perfecting the multidisciplinary treatment, improving the curative effect. Moreover, should be performed molecular biology investigations to study the biological function of markers and lay the biological basis for personalized targeted therapies [2].

Conflicts of Interest: The authors declare no conflict of interest.

References

1. El-Naggar, A.; Chan, J. *WHO Classification of Head and Neck Tumours*, 4th ed.; IARC: Lyon, France, 2017; pp. 232–242.
2. Bajpai, M.; Pardhe, N. Immunohistochemical Expression of CD-177 (c-KIT), P-53 and Ki-67 in Adenoid Cystic Carcinoma of Palate. *J. Coll. Phys. Surg. Park.* **2018**, *28*, S130–S132, doi:10.29271/jcpsp.2018.06.S130.

© 2019 by the authors. Licensee MDPI, Basel, Switzerland. This article is an open access article distributed under the terms and conditions of the Creative Commons Attribution (CC BY) license (http://creativecommons.org/licenses/by/4.0/).

Extended Abstract

Development of a Prognostic Model for Tongue Squamous Cell Carcinoma [†]

Giuseppe Troiano [1],*, Khrystyna Zhurakivska [1], Marco Mascitti [2], Andrea Santarelli [2], Giuseppina Campisi [3] and Lorenzo Lo Muzio [1]

1. Department of Clinical and Experimental Medicine, University of Foggia, 71122 Foggia, Italy; Khrystyna.zhurakivska@unifg.it (K.Z.); lorenzo.lomuio@unifg.it (L.L.M.)
2. Department of Clinical Specialistic and Dental Sciences, Marche Polytechnic, 60131 Ancona, Italy; marcomascitti86@hotmail.com (M.M.); andrea.santarelli@univpm.it (A.S.)
3. Department of Surgical, Oncological and Oral Sciences, University of Palermo, 90127 Palermo, Italy; campisi@odonto.unipa.it
* Correspondence: giuseppe.troiano@unifg.it; Tel.: +39-348-898-6409
† Presented at the XV National and III International Congress of the Italian Society of Oral Pathology and Medicine (SIPMO), Bari, Italy, 17–19 October 2019.

Published: 11 December 2019

One of the objectives of current researches is to be able to customize the treatment of cancer patients. This can be possible only by better stratifying patients based on the most significant prognostic factors [1]. The current staging system for oral cancer based on the 8th edition of American Joint Committee on Cancer (AJCC) [2,3] takes into consideration also depth of invasion and extra-nodal extension (ENE) for patients' stratification [4]. The aim of the present study was to retrospectively evaluate the prognostic value of tumor-stroma ratio in patients with Tongue Squamous Cell Carcinoma (TSCC) and to develop a prognostic model based on the most significant clinical-pathological features. Clinical and pathological data of 211 patients treated for TSCC were collected. 139 patients were re-staged according to the 8th edition of AJCC. Evaluation of TSR was performed on H&E slides and correlation with survival outcomes was evaluated. In particular, disease-specific survival (DSS) and disease-free survival (DFS) were analyzed A prognostic nomogram, based on significantly predictive variables included into a Cox Proportional Hazard model was developed. Low TSR showed to have a negative prognostic value in terms of Disease Specific Survival (DSS) and Overall Survival (OS) for both the 7th and 8th edition classifications. Stage, perineural invasion and Gender significantly correlated to the prognosis of TSCC patients primarily treated by means of surgery. The model built on such parameters showed a good predictive capacity, overperforming the AJCC 8 staging system in stratifying survival in TSCC. The model developed using Gender, TSR and Perineural Invasion and 8th edition of the AJCC staging system could improve TSCC patients' stratification and treatment decisions and represents another step toward the long road for personalized treatment in TSCC patients.

Conflicts of Interest: The authors declare no conflict of interest.

References

1. Almangush, A.; Youssef, O. Does evaluation of tumour budding in diagnostic biopsies have a clinical relevance? A systematic review. *Histopathology* **2019**, *74*, 536–544, doi:10.1111/his.13793.
2. Kowalski, L.P.; Köhler, H.F. Relevant changes in the AJCC 8th edition staging manual for oral cavity cancer and future implications. *Chin. Clin. Oncol.* **2019**, *8*, doi:10.21037/cco.2019.03.01.
3. Moeckelmann, N.; Ebrahimi, A. Prognostic implications of the 8th edition American Joint Committee on Cancer (AJCC) staging system in oral cavity squamous cell carcinoma. *Oral Oncol.* **2018**, *85*, 82–86, doi:10.1016/j.oraloncology.2018.08.013.

4. Mascitti, M.; Rubini, C. American Joint Committee on Cancer staging system 7th edition versus 8th edition: Any improvement for patients with squamous cell carcinoma of the tongue? *Oral Surg. Oral Med. Oral Pathol. Oral Radiol.* **2018**, *126*, 415–423, doi:10.1016/j.oooo.2018.07.052.

© 2019 by the authors. Licensee MDPI, Basel, Switzerland. This article is an open access article distributed under the terms and conditions of the Creative Commons Attribution (CC BY) license (http://creativecommons.org/licenses/by/4.0/).

Extended Abstract

Persistent Idiopathic Facial Pain Associated with Patent Foramen Ovale with Right- to-Left Shunt and Hyperhomocysteinaemia: When a Symptom Can Save a Life †

Daniela Adamo *, Noemi Coppola, Giulio Fortuna, Elena Calabria, Roberto Carbone and Michele D. Mignogna

Department of Neurosciences, Reproductive and Odontostomatological Sciences, University of Naples "Federico II" Via Pansini 5, 80131 Naples, Italy; noemi.coppola91@gmail.com (N.C.); giulio.fortuna@gmail.com (G.F.); calabriaelena92@gmail.com (E.C.); robertocarbone26@gmail.com (R.C.); mignogna@unina.it (M.D.M.)
* Correspondence: danielaadamo.it@gmail.com; Tel.: +393925253864
† Presented at the XV National and III International Congress of the Italian Society of Oral Pathology and Medicine (SIPMO), Bari, Italy, 17–19 October 2019.

Published: 12 December 2019

1. Background

The association of Persistent Idiopathic Facial Pain (PIFP) and a congenital heart abnormality has never previously been reported. We report a case in which the diagnostic workup performed for PIFP revealed a Patent Foramen Ovale (PFO), with right-to-left shunt (RLS) and hyperhomocysteinaemia due to a polymorphism in the gene coding for 5,10 Methylene TetraHydroFolate Reductase (MTHFR) (c677) with a homozygous mutation.

2. Presentation of the Case

A 48-year-old female patient presented at the Oral Medicine Unit with a 10-month history of left-side facial pain. The pain was deep, occurring daily and diffuse to the upper teeth irradiating to the ipsilateral jaw without any paroxysms. She had undergone several investigations and tooth extractions without any relief and had been unsuccessfully treated with amitriptyline, pregabalin and triptan. She had a history of venous thromboembolism occurring at the age of 32 years and never evaluated. Gliosis and carotid siphons tortuosity were found on Magnetic Resonance Imaging (MRI) of the brain (Figure 1a) and hyperhomocysteinaemia (18 μmol/L) was identified. Transcranial Doppler Ultrasonography (TCD) with an agitated saline test was performed and showed a high degree of RLS (Figure 1b) while Transesophageal Echocardiography detected an interatrial septum aneurysm with PFO. Complete thrombophilia screening revealed a homozygosis mutation for MTHFR (c677, Ala--> Val).

A complete remission of pain was reported three months after the percutaneous closure of the PFO and therapy with antiplatelet and folic acid supplementation. After six months the patient continues to stay in well-being only undergoing this treatment without the need for any selective drugs for PIFP.

3. Conclusions

To the best of our knowledge, this is the first case described of facial pain secondary to a RLS due to an asymptomatic PFO in a patient with a prothrombotic state in which the disappearance of pain after the percutaneous closure of the PFO supports the possible association between a RLS and PIFP. PFO with RLS has been suggested as a risk factors for cryptogenic stroke in younger patients.

Therefore, in our case, the early detection could be considered a possible lifesaver. However, the study leaves an unresolved question: namely whether the association is coincidental or causal. To address this point, the mechanism of action has to be further elucidated. This report shows that a careful evaluation of the clinical history, together with an interdisciplinary collaboration and a specific work up including neuroimaging studies is recommended in every case of facial pain in order both to achieve a more accurate and early recognition and to avoid diagnostic delay, which could sometimes be fatal for the patient.

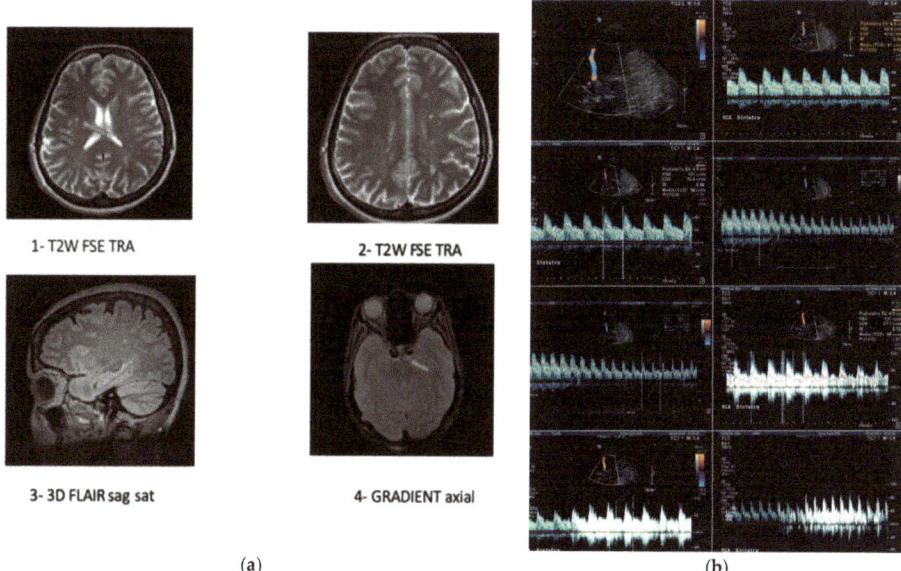

Figure 1. (**a**) Magnetic Resonance Imaging (MRI) of the brain and brainstem with and without intravenous paramagnetic contrast showed: 1–2 T2W TSE FSE TRA show gliotic areas of the white substance of the brain 3 3D FLAIR SAG SAT shows gliotic areas in of the white substance of the brain. 4 GRADIENT AXIAL asymmetry and tortuosity of carotid siphons (**b**) Transcranial Doppler Ultrasonography (TCD) with the agitated saline test showed a high degree of shunt with uncountable microembolic signals (the "curtain effect").

Conflicts of Interest: The authors declare no conflict of interest.

References

1. Kumar, P.; Kijima, Y.; West, B.H.; Tobis, J.M. The Connection Between Patent Foramen Ovale and Migraine. *Neuroimaging Clin. N. Am.* **2019**, *29*, 261–270, doi:10.1016/j.nic.2019.01.006. Review.
2. Agostoni, E.C.; Longoni, M. Migraine and cerebrovascular disease: Still a dangerous connection? *Neurol. Sci.* **2018**, *39* (Suppl. 1), 33–37, doi:10.1007/s10072-018-3429-8.
3. Shi, Y.J.; Lv, J.; Han, X.T.; Luo, G.G. Migraine and percutaneous patent foramen ovale closure: A systematic review and meta-analysis. *BMC Cardiovasc. Disord.* **2017**, *17*, 203, doi:10.1186/s12872-017-0644-9.
4. Spencer, B.T.; Qureshi, Y.; Sommer, R.J. A retrospective review of clopidogrel as primary therapy for migraineurs with right to left shunt lesions. *Cephalalgia.* **2014**, *34*, 933–937.

 © 2019 by the authors. Licensee MDPI, Basel, Switzerland. This article is an open access article distributed under the terms and conditions of the Creative Commons Attribution (CC BY) license (http://creativecommons.org/licenses/by/4.0/).

Extended Abstract

Salivary Metabolic Analysis in Healthy Subjects and Perspectives for Patients with Oral Cancer: Pilot Study and Systematic Review [†]

Rita Antonelli [1,*], **Margherita Eleonora Pezzi** [1], **Maria Vittoria Viani** [1], **Thelma A. Pertinhez** [2,3], **Eleonora Quartieri** [2,3], **Benedetta Ghezzi** [1], **Giacomo Setti** [1], **Paolo Vescovi** [1] **and Marco Meleti** [1]

1. Centro Universitario di Odontoiatria, Department of Medicine and Surgery, University of Parma, Via Gramsci 14, 43126 Parma, Italy; margherita.pezzi@gmail.com (M.E.P.); mariavittoriaviani@gmail.com (M.V.V.); benedetta.ghezzi@unipr.it (B.G.); setti.giacomo@gmail.com (G.S.); paolo.vescovi@unipr.it (P.V.); marco.meleti@unipr.it (M.M.)
2. Department of Medicine and Surgery, University of Parma, Via Volturno, 39, 43121 Parma, Italy; Thelma.pertinhez@unipr.it (T.A.P.); Eleonora.quartieri@unipr.it (E.Q.)
3. Transfusion Medicine Unit, Azienda USL—IRCCS di Reggio Emilia—Viale Umberto I, 50, 43123 Reggio Emilia, Italy
* Correspondence: rita.antonelli@hotmail.it; Tel.: +39-346-969-2224
† Presented at the XV National and III International Congress of the Italian Society of Oral Pathology and Medicine (SIPMO), Bari, Italy, 17–19 October 2019.

Published: 12 December 2019

1. Introduction

Oral squamous-cell carcinoma, the most frequent malignant neoplasm of the oral cavity, has a poor 5 years survival rate.

The identification of specific salivary biomarkers can lead to a reduction of diagnostic delay.

The aims of the present work are:

1. to report the results of a pilot analysis on metabolic salivary composition of 20 healthy subjects.
2. to perform a systematic review designed to answer to the question: "Is there evidence that support the use of salivary metabolomics for diagnosis of OSCC?"

2. Materials and Methods

Pilot study. Twenty healthy subjects (10 males and 10 females) aged between 20 and 25 years were included. Enrolled subjects underwent a thorough dental examination, in order to assess the presence of inflammatory and/or infectious conditions in the oral cavity, which could interfere with the composition and analysis of saliva.

Whole saliva (WS) and blood were collected at Centro Universitario di Odontoiatria in Parma. The metabolic profile of biofluids was determined by proton nuclear magnetic resonance (^{1}H-NMR) in a JEZ 600 MHz ECZ600R spectrometer at the Interdepartmental Center for Measurements (CIM) of the University of Parma.

The systematic review was performed searching Medline, Scopus and Web of Science, using as entry terms a combination of "saliva" or "salivary biomarkers" and "oral carcinoma", "oral cancer", "squamous cell carcinoma", "salivary metabolomics". endpoint of research was generally 2019. The quality of the studies was assessed by two independent reviewers based on the checklist proposed by the National Institute of Health (NIH). The level of evidence was assessed using the Oxford Center for Evidence-Based Medicine (CEMB) classification.

3. Results

The salivary metabolic analysis identified and quantified over 50 metabolites. Acetate was found to be the metabolite with the highest concentration, both in males and females (Figure 1).

Comparing acetate with the second most concentrated metabolite (propionate) the first has a presence 9.5 times higher than the second.

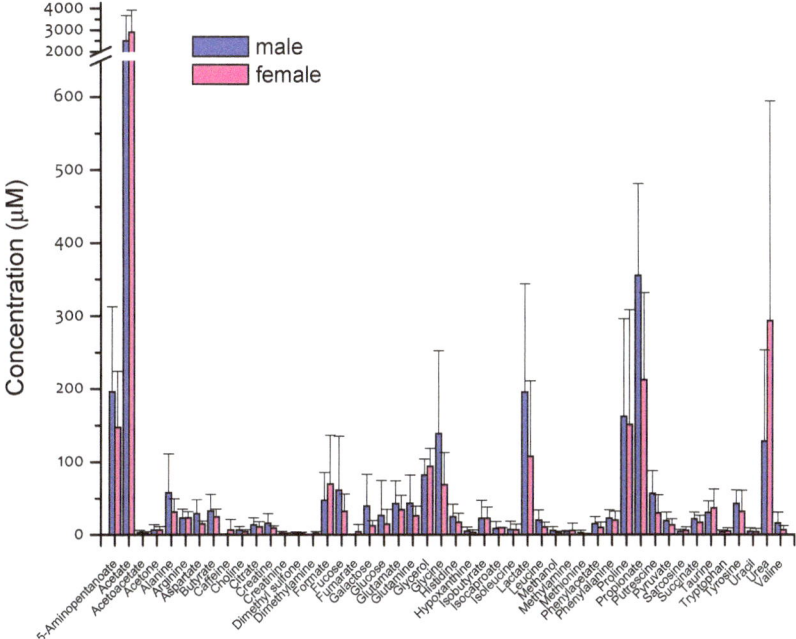

Figure 1. Salivary concentration of identified metabolites divided by gender.

Thirteen papers out of 4198 fulfilled the inclusion criteria. No article obtained a "good" rating (0/13, 0%), 12 articles were classified as "fair" (92%) and 1 article as "poor" (8%).

Most frequent risks of bias were the lack of sample size justification, the absence of concurrent controls, the choice to not blind the status of participants by the investigators.

According to the CEMB classification the 13 articles have a low level of evidence (level 4). Most frequently investigated metabolites that revealed a strong association with OSCC were: pyruvic acid, glycine, proline and choline.

4. Conclusions

Metabolic analysis of saliva is one of the most promising field for early diagnosis oral squamous cell carcinoma.

The present study demonstrates the salivary metabolic composition of healthy subject and may serve as a reference to elucidate saliva alteration found in patient with malignant tumours.

Conflicts of Interest: The authors declare no conflict of interest.

 © 2019 by the authors. Licensee MDPI, Basel, Switzerland. This article is an open access article distributed under the terms and conditions of the Creative Commons Attribution (CC BY) license (http://creativecommons.org/licenses/by/4.0/).

Extended Abstract

Hard Palate Hyperpigmentation Induced by Chloroquine: A Case Report [†]

Sara Attuati [1],*, Valeria Martini [1], Riccardo Bonacina [1], Umberto Mariani [1] and Andrea Gianatti [2]

[1] Department of Dentistry, ASST-PG23: Ospedale Papa Giovanni XXIII, 24100 Bergamo, Italy; vmartini@asst-pg23.it (V.M.); rbonacina@asst-pg23.it (R.B.); umariani@asst-pg23.it (U.M.)

[2] Department of Anatomical Pathology ASST-PG23: Ospedale Papa Giovanni XXIII, 24100 Bergamo, Italy; agianatti@asst-pg23.it

* Correspondence: sara.attuati@gmail.com; Tel.: +39-340-617-9293

† Presented at the XV National and III International Congress of the Italian Society of Oral Pathology and Medicine (SIPMO), Bari, Italy, 17–19 October 2019.

Published: 12 December 2019

1. Introduction

Pigmented oral mucosa lesions can have an exogenous or endogenous nature, depending on the determining causes. Among the exogenous pigmentations are included those related to heavy metal ingestion or intoxication, metal implantation, drug ingestion and "bad habits" like pencil chewing. A physiological and medical anamnesis is important to investigate the etiology of pigmentations [1].

2. Case Report

A 87-years old man was referred to the Oral Pathology division of Papa Giovanni XXIII hospital in Bergamo for xerostomia and temporomandibular joint (TMJ) disorders. The patient's medical history revealed cardiopathy, uncompensated vestibulopathy and seronegative rheumatoid arthritis. He was receiving treatment with Chloroquine (250 mg/die since 2016), Vitamin "D", Cardioaspirin, Atorvastatin and an antihypertensive drug. The patient stated he never smoked.

Intraoral examination showed a flat, homogeneous, grey discoloration regarding the whole hard palate mucosa. The pigmentated area was asymptomatic, symmetric and bilateral (Figure 1).

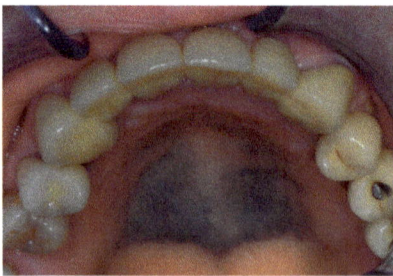

Figure 1. Hard palate hyperpigmentation.

The rest of the oral mucosa appeared to be moistened and healthy, without leucoplasic or eritroplasic areas. Neither functional limitation nor muscular pain was noticed through clinical exam of TMJ.

The patient reported to have never noticed the palate pigmentation before; 15 days later it was still present without any clinical change. After the patient's written informed consent an incisional biopsy was performed in local anesthesia; beside standard histopathologic examination, Prussian blue (Perls') reaction was used to investigate hemosiderin staining (Figure 2).

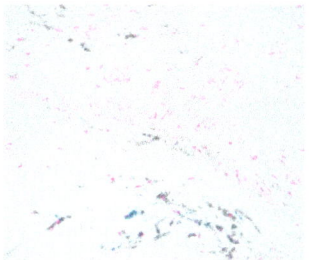

Figure 2. Perls' Prussian blue stained section.

The clinical and histopathologic features confirmed the diagnosis of drug-induced oral pigmentation caused by chloroquine.

3. Discussion

Oral mucosa hyperpigmentation is a rare side effect of antimalarial agents like chloroquine, used also for management of rheumatoid arthritis.

The literature reported that prolonged systemic administration of this drug can determine the deposition of hemosiderin in the tissues of the oral cavity. Chloroquine chelates iron and/or melanin; moreover, melanocytes are stimulated in melanin production [2,3].

In most cases only hard palate is involved, but involvement of gingiva and labial or buccal mucosa has been reported too. For this type of oral pigmentation there is no a recommended treatment, the importance of an early diagnosis of chloroquine-induced hyperpigmentation of the oral mucosa is that it could be a marker of another adverse reaction: irreversible retinopathy. The drug binds pigmented ocular tissues and could lead to blindness. The patient should be referred to an ophthalmologist and, according to the rheumatologist, the drug should be discontinued or modified in the dosage.

Conflicts of Interest: The authors declare no conflict of interest.

References

1. Joseph, A.; Regezi, J.S.; Richard, C.K.; Jordan, A. *Oral Pathology: Clinical Pathologic Correlations*, 7th ed.; Elsevier-Saunders: Amsterdam, The Netherlands, 2017.
2. De Andrade, B.A.B.; Padron-Alvarado, N.A.; Muñoz-Campos, E.M.; Morais, T.M.L.; Martinez-Pedraza, R. Hyperpigmentation of hard palate induced by chloroquine therapy. *J. Clin. Exp. Dent.* **2017**, *9*, e1487–e1491.
3. Tosios, K.I.; Kalogirou, E.M.; Sklavounou, A. Drug-associated hyperpigmentation of the oral mucosa: report of four cases. *Oral Surg. Oral Med. Oral Pathol. Oral Radiol.* **2018**, *125*, e54–e66.

© 2019 by the authors. Licensee MDPI, Basel, Switzerland. This article is an open access article distributed under the terms and conditions of the Creative Commons Attribution (CC BY) license (http://creativecommons.org/licenses/by/4.0/).

Extended Abstract

Plasma Cell Mucositis: A Case Report of an Uncommon Benign Disease [†]

Alessandro Antonelli [1], Fiorella Averta [2], Federica Diodati [2], Danila Muraca [2], Ylenia Brancaccio [1], Michele Davide Mignogna [3] and Amerigo Giudice [2],*

- [1] Department of Health Sciences, University Magna Graecia of Catanzaro, 88100 Catanzaro, Italy; antonellicz@gmail.com (A.A.); ybrancaccio@gmail.com (Y.B.)
- [2] Department of Health Sciences, School of Dentistry, Magna Graecia University of Catanzaro, 88100 Catanzaro, Italy; fiore.averta@gmail.com (F.A.); federicadiodati@libero.it (F.D.); danila.muraca@studenti.unicz.it (D.M.)
- [3] Department of Neurosciences, Reproductive and Odontostomatological Sciences, Federico II University of Naples, 80138 Naples, Italy; mignogna@unina.it
- * Correspondence: a.giudice@unicz.it
- [†] Presented at the XV National and III International Congress of the Italian Society of Oral Pathology and Medicine (SIPMO), Bari, Italy, 17–19 October 2019.

Published: 12 December 2019

Plasma cell mucositis (PCM) is an unusual plasma cell proliferative disorder of the upper aerodigestive tract. It is a rare disease and its etiology is not yet known, it is considered a benign condition of adults and there is no correlation in literature with the development of plasma cell neoplasm. Clinical features are an intensely erythematous mucosa with papillomatous, cobblestone, nodular, or velvety surface changes. Symptoms include dysphagia, oral pain and pharyngitis [1]. Generally, PCM patients have a previous history of autoimmune or immunologically mediated disease. Despite plasma cell mucositis often involves the oral and genital mucosa, there was no genital involvement in this circumstance. We described a case of PCM involving the tongue of a 43-year-old-woman, the patient was referred to the Oral Pathology at the Faculty of Dentistry, Magna Graecia University of Catanzaro, in August 2019. Her past medical history was unremarkable. She reported a burning sensation in her mouth and local dysgeusia on the tip and the right lingual border for 5 years (Figure 1). In the first place several diagnostic hypotheses were proposed, the most of them discarded for incompatibility with blood and laboratory tests. The patient underwent an incisional biopsy under local anesthesia. The specimen was stored in a tube containing formalin 10% and sent to a laboratory for histopathological analysis. Microscopically, a large area of ulceration of the coating epithelium subtended by dense plasma cell infiltrate was observed. The final histopathological diagnosis was "Plasmacytosis of the mucous membranes with restriction for the kappa chains". This disease rarely manifests itself on the tongue, especially in young patients with no comorbidities. The management of PCM is mainly aimed at reducing the symptoms, in fact the patient was treated with systemic prednisone 50 mg/day [2]. Many therapeutic treatments are able to stabilize but are not able to induce a complete remission. PCM is considered an uncommon benign disorder with a favorable prognosis. It is important to differentiate the PCM disease from other neoplastic conditions in order to achieve a better clinical management of the patients.

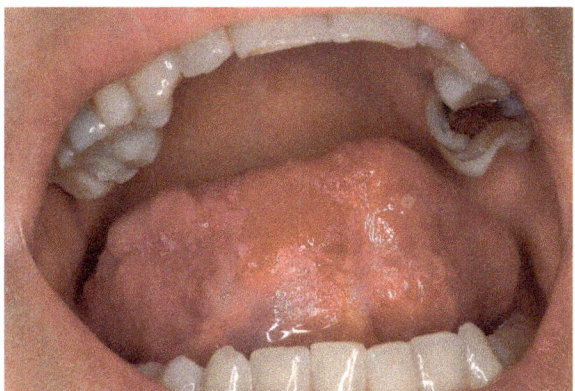

Figure 1. Ulcer localized on the tip of the tongue.

References

1. Solomon, L.W.; Wein, R.O.; Rosenwald, I.; Laver, N. Plasmacellmucositis of the oralcavity: Report of a case and review of the literature. *Oral Surg. Oral Med. Oral Pathol. Oral Radiol. Endod.* **2008**, *106*, 853–860.
2. Gasparro, R.; Adamo, D.; Masucci, M.; Sammartino, G.; Mignogna, M.D. Use of injectable platelet-rich fibrin in the treatment of plasma cell mucositis of the oral cavity refractory to corticosteroid therapy: A case report. *Dermatol. Ther.* **2019**, *32*, e13062.

 © 2019 by the authors. Licensee MDPI, Basel, Switzerland. This article is an open access article distributed under the terms and conditions of the Creative Commons Attribution (CC BY) license (http://creativecommons.org/licenses/by/4.0/).

Extended Abstract

Use of Optical Coherence Tomography in Patients with Desquamative Gingivitis: A Case Series [†]

Vera Panzarella [1,2,*], Alessia Bartolone [2], Domenico Ciavarella [3], Andrea Santarelli [4], Olga Di Fede [2], Rodolfo Mauceri [1] and Giuseppina Campisi [1,2]

1. Unit of Oral Medicine, A.O.U.P "Paolo Giaccone", 90127 Palermo, Italy; rodolfo.mauceri@unipa.it (R.M.); giuseppina.campisi@unipa.it (G.C.)
2. Department of Surgical, Oncological and Oral Sciences (Di.Chir.On.S.), University of Palermo, 90127 Palermo, Italy; alessia.bartolone@community.unipa.it (A.B.); olga.difede@unipa.it (O.D.F.)
3. Department of Clinical and Experimental Medicine, University of Foggia, 71122 Foggia, Italy; domenico.ciavarella@unifg.it
4. Department of Clinical Specialistic and Dental Sciences, Marche Polytechnic University, 60121 Ancona, Italy; andrea.santarelli@staff.univpm.it
* Correspondence: panzarella@odonto.unipa.it
† Presented at the XV National and III International Congress of the Italian Society of Oral Pathology and Medicine (SIPMO), Bari, Italy, 17–19 October 2019.

Published: 12 December 2019

1. Introduction

Desquamative gingivitis (DG) is a descriptive term indicating the presence of erythematous, erosive, desquamative and vesiculo-bullous lesions in the free/attached gingiva, usually expression of several chronic systemic conditions [1].

Optical Coherence Tomography (OCT) is a new biomedical technique based on the interaction of the infrared radiation (900–1500 nm) to the human tissues, allowing the visualization at high resolution of the micro-structural morphology. OCT is proposed in dermatology, ophthalmology and recently in oral medicine [2]. It could be recommended for managing the patients with chronic disease, such as DG.

The aim of this paper is to report a case series of three patients affected by clinical DG valuated by OCT before diagnostic biopsy in order to assess the morphology of the lesions and to guide the oral medicine specialist to hypothetical differential diagnosis.

2. Case Series

Three female patients (mean age 74 ± 9.2 years), with clinical DG, were consecutively recruited at unit of Oral Medicine, and after informed consent, underwent to in vivo OCT examination. In details, we used VivoSight® OCT (Michelson Diagnosis) which is equipped with a flexible fiber optic probe. However, the probe, because it is used for dermatological purposes, is not of the optimal size for the oral cavity. In this way OCT allows to obtains a scan section with width of 6mm and focal depth of 2 mm.

After photo record, OCT examination was performed on each patients in order to identify the most suggestive DG site for the incisional punch biopsy, successively performed by a punches 6 mm diameter (the same size area of OCT scans).

Case 1–2: For both patients, DGs were characterized by erythematous, desquamative, hyperkeratotic lesions. In both cases Nikolsky's sign was negative (Figure 1a).

OCT scans showed: hyperkeratinization, unhomogeneity and decrease of the epithelial layers and increased sub-epithelial cellularity under basement-membrane (BM) (Figure 1b).

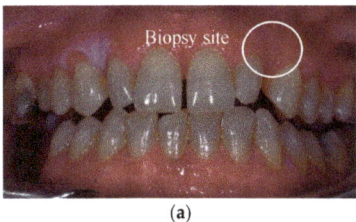 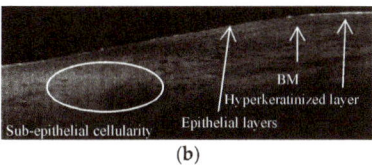

Figure 1. (Case 1) (a) Atrophic-erosive, desquamative and hyperkeratotic DG, (biopsy site underlined); (b) OCT scan of the same site of punch biopsy, with morphological details.

Case 3: DG showed erythematous, desquamative, hyperkeratotic areas, associated with a positive Nikolsky's sign (Figure 2a).

OCT scan showed a compact epithelial layer entirely separated from the BM by interposed fluid that appears dark, and potentially suggestive of a sub-epithelial blister (Figure 2b) [3].

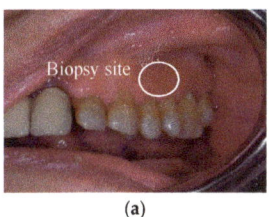 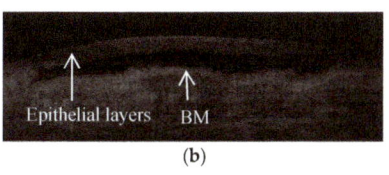

Figure 2. (Case 3) (a) Atrophic-erosive, hyperkeratotic and vesicular DG, (biopsy site underlined); (b) OCT scan shows evident detachment of the epithelium from BM zone.

Histological reports of **Case 1** and **2** were Oral Lichen Planus; whereas **Case 3** was Mucous Membrane Pemphigoid (MMP).

3. Conclusions

We presented three patients with similar clinical-features of DGs, that showed a morphological diversity by OCT investigation.

In details, in **case 1–2** were present an alteration of epithelial layers with a greater sub-epithelial cellularity, reported in OCT scans could be evocative of a chronic inflammatory infiltrate, usually histologically described in OLP.

In **case 3** it was a clear evidence of a sub-epithelial blister that could early guider the clinician to a diagnosis sub-epithelial vesicular-bullous pathologies (I.E MMP).

The results of our study, although reported to a small simple size, prompt a potential use of OCT in clinical management of DG. In particular, for preliminary assessment of the lesions, in order to suggest a more specific sites for histological investigation.

Conflicts of Interest: The authors declare no conflict of interest.

References

1. Lo Russo, L.; Fierro, G.; Guiglia, R.; Compilato, D.; Testa, N.F.; Lo Muzio, L.; Campisi, G. Epidemiology of desquamative gingivitis: Evaluation of 125 patients and review of the literature. *Int. J. Dermatol.* **2009**, *48*, 1049–1052. doi:10.1111/j.1365-4632.2009.04142.x.
2. Huang, D.; Swanson, E.A.; Lin, C.P.; Schuman, J.S.; Stinson, W.G.; Chang, W.; Hee, M.R.; Flotte, T.; Gregory, K.; Puliafito, C.A.; et al. Optical coherence tomography. *Science* **1991**, *254*, 1178–1181.

3. Capocasale, G.; Panzarella, V.; Rodolico, V.; Di Fede, O.; Campisi, G. In vivo optical coherence tomography imaging in a case of mucous membrane pemphigoid and a negative Nikolsky's sign. *J. Dermatol.* **2018**, *45*, 603–605. doi:10.1111/1346-8138.14267.

 © 2019 by the authors. Licensee MDPI, Basel, Switzerland. This article is an open access article distributed under the terms and conditions of the Creative Commons Attribution (CC BY) license (http://creativecommons.org/licenses/by/4.0/).

Extended Abstract

Cartilaginous Choristoma of the Lower Lip †

Moreno Bosotti [1,*], Francesca Boggio [2], Anna Mascellaro [3], Margherita Rossi [1], Massimo Porrini [1], Ettore del Rosso [3] and Francesco Spadari [1]

1. Division of Oral Pathology and Oral Medicine, Department of Biomedical, Surgical and Dental Sciences University of Milan, Maxillo-Facial and Odontostomatology Unit, Cà Granda University Hospital, IRCCS Foundation, 20122 Milan, Italy; margherita.rossi2@unimi.it (M.R.); massimo.porrini@unimi.it (M.P.); francesco.spadari@unimi.it (F.S.)
2. Division of Pathology, Department of Biomedical, Surgical and Dental Sciences University of Milan, Maxillo-Facial and Odontostomatology Unit, Cà Granda University Hospital, IRCCS Foundation, 20122 Milan, Italy; francesca.boggio@policlinico.mi.it
3. Division of Oral Surgery; Department of Biomedical, Surgical and Dental Sciences University of Milan, Maxillo-Facial and Odontostomatology Unit, Cà Granda University Hospital, IRCCS Foundation, 20122 Milan, Italy; anna.mascellaro@unimi.it (A.M.); delrosso.ettore@unimi.it (E.d.R.)

* Correspondence: moreno.bosotti@unimi.it; Tel.: +39-340-8240550
† Presented at the XV National and III International Congress of the Italian Society of Oral Pathology and Medicine (SIPMO), Bari, Italy, 17–19 October 2019.

Published: 12 December 2019

1. Introduction

The term choristoma applies to cohesive tumor-like mass composed by histologically normal tissue in abnormal locations. Cartilaginous choristomas are a rare finding in the mouth where most frequently arise in the tongue and less commonly in sites such as the buccal mucosa, soft palate and gingiva.

In literature cartilaginous choristomas of the lower lip have been described in 3 women [1].

We report a case of a cartilaginous choristoma occurring in the lower lip in a young man.

2. Material and Methods

A caucasian 37-year-old man referred to our division complaining an asymptomatic and slow-growing lesion at the lower lip. To the patient's knowledge, the lesion was present since 12 months. There was no reported history of direct trauma or infection in that area before the appearance of the swelling. On examination, there was asymptomatic, mobile submucosal area within the lower lip. Due to the presentation and location, several entities were considered in the differential diagnosis: traumatic fibroma, mucocele, inflamed minor salivary gland, lipoma or a minor salivary gland tumour. In order to obtain the correct diagnostic interpretation, the lesion was then surgically excised.

3. Discussion

The exact cause of a cartilaginous choristoma is unknown. Several theories have been proposed to explain the origin of the cartilage in the soft tissues of oral cavity. However, the embryonic theory and the metaplastic theory are the two main theories favored in the literature. The first theory proposes that the lesions arises from heterotopic fetal cartilaginous remnants.

The metaplastic theory suggests instead that cartilaginous choristomas develop from pluripotent mesenchymal cells either de novo or potentially stimulated by trauma, irritation or chronic inflammation [2].

On gross examination the lesion excised appeared as a nodular formation of 1.3 cm in major dimension, covered by flat and congested mucosa. In section, this formation showed a white-yellowish, focally translucent, surface and soft-elastic consistency.

Microscopically the lesion consisted in a non-capsulated nodule of the sub-epithelial chorion, with well demarcated margins, composed by mature chondroid tissue admixed with fibrous and mature adipose tissue (Figure 1). There was no significant cytological atypia, neither necrosis or evident mitosis. The overlying mucosal layer was uninvolved without significant histological alterations. A mild and scattered lymphoid infiltrate was also associated.

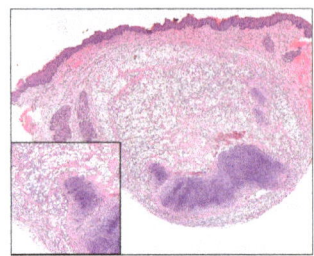

Figure 1. Hematoxylin-Eosin 5× (Insert: Hematoxylin-Eosin 20×). Non-capsulated sub-epithelial formation consisting in mature chondroid tissue with fibro-adipose component without any atypia or necrosis.

4. Conclusions

This case report depicts a cartilaginous choristoma developed in a rare site.

Unlike the cartilaginous choristomas involving the tongue, where the prevalence is similar among males and females, the lip lesions, albeit reported only in a small number of cases, seems to affect mostly female subjects with a female/male ratio of 3:1.

To our knowledge, this is the first case of a cartilaginous choristoma of the lower lip presenting in a male described in literature.

Conflicts of Interest: The authors declare no conflict of interest.

References

1. Halley, D.; Dargue, A.; Pring, M. Cartilaginous Choristoma of the lower lip: Report of a case and review of the literature. *Oral Surg.* **2014**, *7*, 48–50.
2. Batra, R. The pathogenesis of oral choristomas. *J. Oral and Maxillofac. Surg. Med. Pathol.* **2012**, *24*, 110–114.

© 2019 by the authors. Licensee MDPI, Basel, Switzerland. This article is an open access article distributed under the terms and conditions of the Creative Commons Attribution (CC BY) license (http://creativecommons.org/licenses/by/4.0/).

Extended Abstract

Describing Clinical and Histological Outcome of Oral Cancer Patients with Recurrent Malignant or Premalignant Oral Lesions: A Retrospective Series with a Follow-Up of 15 Years [†]

Adriana Cafaro *, Marco Cabras, Alessio Gambino, Marco Garrone, Paolo Giacomo Arduino and Roberto Broccoletti

Department of Surgical Sciences, C.I.R.—Dental School, Oral Medicine section, University of Turin, 10126 Turin, Italy; cabrasmarco300@gmail.com (M.C.); alessio.gambino@unito.it (A.G.); ma.garrone@libero.it (M.G.); paologiacomo.arduino@unito.it (P.G.A.); roberto.broccoletti@unito.it (R.B.)
* Correspondence: adri.cafaro@gmail.com; Tel.: +39-011-633-15-22
† Presented at the XV National and III International Congress of the Italian Society of Oral Pathology and Medicine (SIPMO), Bari, Italy, 17–19 October 2019.

Published: 12 December 2019

1. Introduction

Treatment of oral potentially malignant disorders (OPMD) and oral malignant lesions does not simply ends with the complete removal of the affected tissues. Since then, the oral physician plays a key role in terms of secondary prevention, carried out through an appropriate guidance of the patient to a change of lifestyle, and regular post-therapy surveillance. Relapse for head and neck squamous cell carcinoma varies from 16% to 52% [2]. Patients with primary tumor diagnosis also have a higher chance to develop second primary tumors (SPTs), arising further than 2 cm from the primary site and/or 5 years later than primary diagnosis, due to an independent carcinogenesis leading to the onset of a new tumor [1,2]. Aim of the present study was to review retrospectively, from 2003 to 2018, how many patients, firstly diagnosed with an oral malignant neoplasm, experienced a subsequent malignant and/or premalignant disorder, in the form of new OPMD, cancerized OPMD, relapsed cancer or SPT.

2. Methods

Electronic archive of the Oral Medicine Section of University of Turin, Italy was reviewed. Inclusion criteria were the following: patients firstly diagnosed with a malignant neoplasm, who later developed at least one OPMD or another malignant neoplasm. The following data were acquired: histological diagnosis, site of occurrence, new OPMD, relapsed OPMD, SPTs and relapsed cancer.

3. Results

A sample of 79 patients (41 F, 38 M; mean age 67.5 years) was finally examined, associated with 199 histological reports. Figures 1 and 2 show their distribution based on histology and anatomic site.

Of the 199 lesions detected, 111 (55.8%) were premalignant, whereas 88 (44.2%) where malignant. Seventy-two of the 88 malignancies (81.8%) were diagnosed as oral squamous cell carcinoma (OSCC). Of 199 lesions, 70 could be categorized as relapsed lesions: 38 were diagnosed as OPMD and 32 as neoplasms. Of the remaining 129 new lesions, either detected 2 cm further from the primary site or 5 years after diagnosis, 62 were SPT, 6 severe OPMD with SPT-like features, and 61 OPMD.

4. Discussion

In this case-series, tongue was by far the most affected site for premalignant and malignant recurrent disorders. Rate of oral cancer relapse in our series aligned with that of the most recent literature (36.3% vs. range of 14–52% by Netto and co-workers) [2].

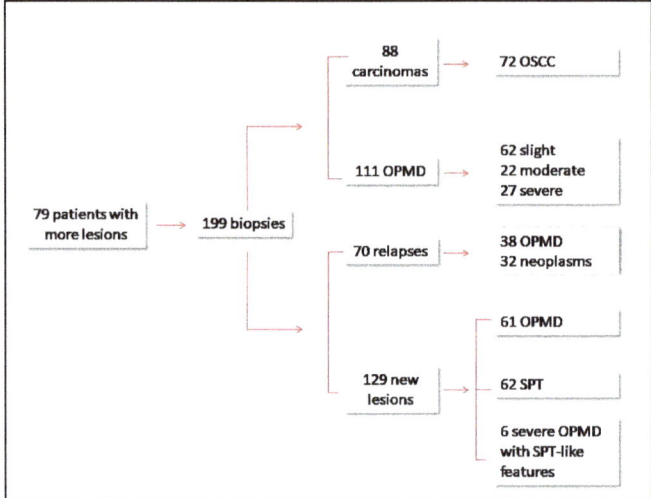

Figure 1. Flow-chart showing the distribution for type of lesion and new lesions/relapses ratio.

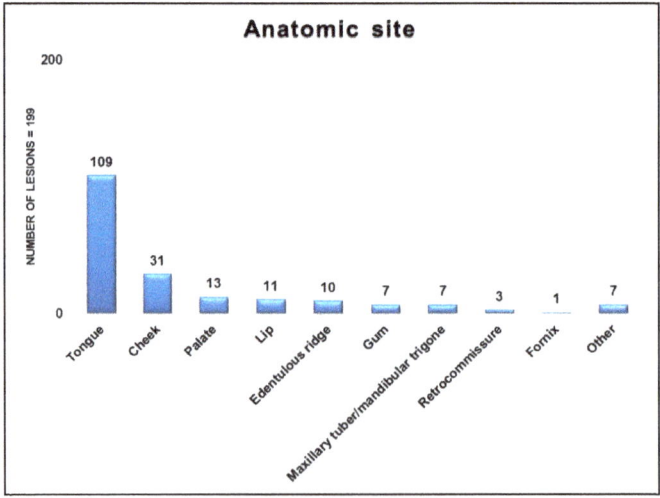

Figure 2. Distribution of the 199 lesions according to the anatomic site.

References

1. Brands, M.T.; Smeekens, E.A.J.; Takes, R.P.; Kaanders, J.H.A.M.; Verbeek, A.L.M.; Merkx, M.A.W.; Geurts, S.M.E. Time patterns of recurrence and second primary tumors in a large cohort of patients treated for oral cavity cancer. *Cancer Med.* **2019**, *8*, 5810–5819.
2. Gleber-Netto, F.O.; Braakhuis, B.J.; Triantafyllou, A.; Takes, R.P.; Kelner, N.; Rodrigo, J.P.; Strojan, P.; Vander Poorten, V.; Rapidis, A.D.; Rinaldo, A.; et al. Molecular events in relapsed oral squamous cell carcinoma: Recurrence vs. secondary primary tumor. *Oral Oncol.* **2015**, *51*, 738–744.

© 2019 by the authors. Licensee MDPI, Basel, Switzerland. This article is an open access article distributed under the terms and conditions of the Creative Commons Attribution (CC BY) license (http://creativecommons.org/licenses/by/4.0/).

Extended Abstract

Management of Oral Hydroxyurea-Related Ulcers: A Cases Series †

Davide Conrotto [1,*], Paolo G. Arduino [1], Roberto Freilone [2], Paola Carcieri [1], Alessio Gambino [1] and Roberto Broccoletti [1]

[1] Department of Surgical Sciences, Oral Medicine Section, CIR-Dental School, University of Turin, 10126 Turin, Italy; paologiacomo.arduino@unito.it (P.G.A.); carcieri-paola@libero.it (P.C.); alegam33@hotmail.it (A.G.); roberto.broccoletti@unito.it (R.B.)

[2] Department Structure of Hematology, ASLTO4, 10015 Ivrea (To), Italy; rfreilone@aslto4.piemonte.it

* Correspondence: davide.conrotto@gmail.com; Tel.: +39-11-6331522

† Presented at the XV National and III International Congress of the Italian Society of Oral Pathology and Medicine (SIPMO), Bari, Italy, 17–19 October 2019.

Published: 12 December 2019

1. Introduction

Hydroxyurea (HU) is an anti-cancer agent commonly used in myeloproliferative Philadelphia negative disorders as polycythemia vera (PV), essential thrombocitosis (ET) and myelofibrosis (MF). Most frequent side effects of HU are skin numbness or purple discoloration, skin ulcers or open sores and low blood cell counts. Using hydroxyurea may increase risk of developing other types of cancer, such as leukemia or skin cancer. Some recent studies underline the possibility of an oral toxicity, but its frequency seems to be quite rare. We present a cases series in which the oral lesions happened in patients in treatment with HU, focusing on their management.

2. Methods

A total of 7 patients (3 males, 4 females, mean age 71), arrived from January 2018 to June 2019 to the Oral Medicine section of the Dental school Lingotto, University of Turin, were enrolled in this study. 4 of them were affected by PV, 1 by MF, 2 by ET. Patients were invited to an oral visit by their hematologist. The visit was performed by a doctor expert in oral medicine. When the diagnosis of oral mucosa ulcers was done, the first line treatment was chlorexihidine mouth rinses and a topical cortisteroid applied 3 times daily for 2 weeks (clobetasol 0.05%). Only if this treatment failed or a serious new episode happened, a reduction of HU dose or its suspension were evaluated.

3. Results

Results are summarized in Table 1.

Table 1. Characteristics of patients and treatments strategies.

Pz	Age	Sex	Haematological Disease	HU Therapy (Years)	HUDose (g/Week)	Jak2	Treatment
1	80	f	PV	3	7	pos	Clobetasol, HU suspension
2	70	m	PV	6	9	pos	Clobetasol
3	70	f	PV	5	7	pos	Clobetasol, HU riduction
4	88	m	MF	5	3	pos	Clobetasol
5	64	m	PV	14	9	pos	Clobetasol, HU suspension
6	61	f	ET	2	6	neg	Clobetasol
7	64	f	ET	2	9	neg	Clobetasol

PV = polycythemia vera, ET = essential thrombocytosis, MF = myelofibrosis HU = hydroxyurea.

The clinical presentation of oral lesions was aphthous-like ulcers. None of the patients had a previous history of recurrent aphthous stomatitis. Oral toxicity appeared after a mean period of 5.2 years. A complete resolution of oral ulcers was observed in 4/7 patients (57.2%), 3 patients had no response or new episode that lead to suspension of HU (2 pz, 28.5%) or its dose reduction (1 pz, 14.2%).

4. Discussion

This condition is considered an early complication, mainly within the first year [1,2]. Previous studied reported that from 0.8 to1.8% of all patients treated with HU developed oral aphthous ulcers [2,3]. Our study, with a medium period of 5.2 years of HU, presents a later onset of the oral toxicity. Other authors [3] suggest a therapy based on mouthwashes with folic acid and vitamin A. In our practice, topical corticosteroids and chlorhexidine mouth rinses seem to be a good treatment for oral lesions, but, when failed, the reduction of the HU dose or its suspension are the only way to solve the problem. Good prospective could arrive from the new cytostatic agents (e.g., ruxolitib, anagrelide).

Conflicts of Interest: The authors declare no conflict of interest.

References

1. Badawi, M.; Almazrooa, S.; Azher, F.; Alsayes, F. Hydroxyurea-induced oral ulceration. *Oral Surg. Oral Med. Oral Pathol. Oral Radiol.* **2015**, *120*, e232–e234.
2. Latagliata, R.; Spadea, A.; Cedrone, M.; Di Giandomenico, J.; De Muro, M.; Villivà, N.; Breccia, M.; Anaclerico, B.; Porrini, R.; Spirito, F.; et al. Symptomatic mucocutaneous toxicity of hydroxyurea in Philadelphia chromosome-negative myeloproliferative neoplasms: The Mister Hyde face of a safe drug. *Cancer* **2012**, *118*, 404–409.
3. Antonioli, E.; Guglielmelli, P.; Pieri, L.; Finazzi, M.; Rumi, E.; Martinelli, V.; Vianelli, N.; Luigia Randi, M.; Bertozzi, I.; De Stefano, V.; et al. Hydroxyurea-related toxicity in 3411 patients with Ph'-negative MPN. *Am. J. Hematol.* **2012**, *87*, 552–554.

© 2019 by the authors. Licensee MDPI, Basel, Switzerland. This article is an open access article distributed under the terms and conditions of the Creative Commons Attribution (CC BY) license (http://creativecommons.org/licenses/by/4.0/).

 *proceedings*

Extended Abstract

Odontostomatological Findings in Heimler Syndrome: A Case Report [†]

Antonio Romano [1], Maria Rosaria Barillari [2], Carlo Lajolo [3], Fedora della Vella [4], Giuseppe Costa [2], Alberta Lucchese [1], Rosario Serpico [1], Francesca Simonelli [1] and Maria Contaldo [1,*]

1. Multidisciplinary Department of Medical-Surgical and Odontostomatological Specialties, University of Campania "L. Vanvitelli", 80138 Naples, Italy; antonio.romano4@unicampania.it (A.R.); alberta.lucchese@unicampania.it (A.L.); rosario.serpico@unicampania.it (R.S.); francesca.simonelli@unicampania.it (F.S.)
2. Department of Mental and Physical Health and Preventive Medicine, University of L. Vanvitelli, 80138 Naples, Italy; mariarosaria.barillari@unicampania.it (M.R.B.); giuseppe.costa@unicampania.it (G.C.)
3. Head and Neck Department, Fondazione Policlinico Universitario A. Gemelli-IRCCS, School of Dentistry, Università Cattolica del Sacro Cuore, 00168 Rome, Italy; clajolo@gmail.com
4. Interdisciplinary Department of Medicine, University "Aldo Moro" of Bari, 70121 Bari, Italy; dellavellaf@gmail.com
* Correspondence: maria.contaldo@gmail.com or maria.contaldo@unicampania.it; Tel.: +39-3204876058
† Presented at the XV National and III International Congress of the Italian Society of Oral Pathology and Medicine (SIPMO), Bari, Italy, 17–19 October 2019.

Published: 12 December 2019

Heimler syndrome (HS) is rare autosomal-recessive disorder caused by mutations of peroxin genes, *PEX1* and *PEX6*. Defects in peroxin genes alter peroxisome assembly and its metabolic pathways, essential for the metabolism of fatty acids, ether lipids, polyamines and amino acids, thus supporting the peroxisome biogenesis disorders (PBDs): a variety of severe conditions, among which HS is the mildest form. Heimler et al. first described HS in 1991 [1], and, since then, scientific literature has reported less than 12 families and less than 20 cases. HS is generally characterized by pre-lingual hearing loss, nail abnormalities, ocular involvements and dental anomalies [2]. Here we report a case of a 9-year-old female, whose genetic analyses revealed to be affected by HS and who underwent multi-disciplinary examinations to define her complete clinical features. She referred at the Eye clinic of the University of Campania "L. Vanvitelli" for ophthalmological evaluations. At the beginning, the presence of sensorineural hearing loss and early onset atypical Retinitis Pigmentosa, oriented toward the clinical diagnosis of Usher syndrome (US), another autosomal-recessive disorder, typically characterized by partial/total hearing loss and worsening vision loss. Hence, molecular genetic tests were performed to refine the diagnosis. After sequencing more than 2000 genes, it results two pathogenic variants of PEX1, thus excluding US and orienting toward HS. At this point, a multi-disciplinary team approached the patient to identify the further HS features. In addition to ophthalmological and audiological findings, clinical and instrumental phoniatric and neuropsychological examinations revealed deficit in language skills and moderate intellectual disability; dermatologists confirmed the presence of leukonychia. Odontostomatological evaluation was performed by two oral pathologists expert in pediatric dentistry. After excluding facial disharmonies at the extraoral examination, the intraoral one revealed a mixed dentition, as expecting according to age [3]. The upper central incisors displayed white spots on the vestibular surfaces and the first permanent molars cusps were yellowish and hypoplastic. The X-ray orthopantomography revealed no agenesis and a pale reduction of the enamel density of canines and premolars, whose cusps appeared slightly hypoplastic. The patient reported mouth breathing, probably responsible for the anterior open bite and the high-arched palate. Oral mucosae were healthy and normochromic. Tongue and teeth were covered by a visible layer of dental plaque, which has been removed by professional oral hygiene. To date the patient is under multidisciplinary follow-up for her various affections, included the dental ones. It is notably the unanimity of the literature in reporting dental defects in HS. Which differs among the reports is the ascription of these defects under the diction of "amelogenesis imperfecta" given by some authors, despite the lack of genetic tests supporting defective genes involved in amelogenesis. Hence, we retain the condition here presented closer to the molar incisor hypomineralization (MIH), whose causes are still not clearly defined [4].

References

1. Heimler, A.; Fox, J.E.; Hershey, J.E.; Crespi, P. Sensorineural hearing loss, enamel hypoplasia, and nail abnormalities in sibs. *Am. J. Med. Genet.* **1991**, *39*, 192–195.
2. Contaldo, M.; Di Stasio, D.; Santoro, R.; Laino, L.; Perillo, L.; Petruzzi, M.; Lauritano, D.; Serpico, R.; Lucchese, A. Non-invasive in vivo visualization of enamel defects by reflectance confocal microscopy (RCM). *Odontology* **2015**, *103*, 177–184, doi:10.1007/s10266-014-0155-4.
3. Gentile, E.; Di Stasio, D.; Santoro, R.; Contaldo, M.; Salerno, C.; Serpico, R.; Lucchese, A. In vivo microstructural analysis of enamel in permanent and deciduous teeth. *Ultrastruct. Pathol.* **2015**, *39*, 131–134, doi:10.3109/01913123.2014.960544.
4. Contaldo, M.; Serpico, R.; Lucchese, A. In vivo imaging of enamel by reflectance confocal microscopy (RCM): Non-invasive analysis of dental surface. *Odontology* **2014**, *102*, 325–329, doi:10.1007/s10266-013-0110-9.

© 2019 by the authors. Licensee MDPI, Basel, Switzerland. This article is an open access article distributed under the terms and conditions of the Creative Commons Attribution (CC BY) license (http://creativecommons.org/licenses/by/4.0/).

Extended Abstract

Lacosamide in the Treatment of Trigeminal Neuralgia Refractory to Conventional Treatment Due to Severe Leukopenia Induced by Anticonvulsants [†]

Noemi Coppola *, Elena Calabria, Giulio Fortuna, Elvira Ruoppo, Marco Caparrotti and Daniela Adamo

Department of Neuroscience, Reproductive and Odontostomatological Sciences, University of Naples Federico II. Via Pansini 5, 80131 Naples, Italy; calabriaelena92@gmail.com (E.C.); giulio.fortuna@gmail.com (G.F.); elviraruoppo@gmail.com (E.R.); marcocaparrotti@virgilio.it (M.C.); danielaadamo.it@gmail.com (D.A.)
* Correspondence: noemi.coppola91@gmail.com; Tel.: +393392602615
† Presented at the XV National and III International Congress of the Italian Society of Oral Pathology and Medicine (SIPMO), Bari, Italy, 17–19 October 2019.

Published: 12 December 2019

1. Background

Trigeminal Neuralgia (TN) is one of the most frequent and severe types of neuropathic facial pain encountered by clinicians [1]. The first line medical approach generally involves treatment with first and second generation anticonvulsants [2]. However, unfavorable adverse events impair compliance with the treatment, especially in relation to elderly patients. Lacosamide (LCM) is a third-generation anticonvulsant drug; it is a functionalized amino acid, with a multimodal mechanism of action which is not completely clear (Table 1 and Figure 1) [3,4]. It has also come into consideration as a treatment for neuropathic pain.

We have aimed to investigate the efficacy and safety of LCM as a monotherapy in a case of TN who had not responded positively to previous treatments on account of severe leukopenia induced by several of the anticonvulsants previously used.

Table 1. Mechanism of action and advantages of treatment with Lacosamide.

Mechanism of Action of Lacosamide
Enhancing slow inactivation of VGSC peripheral (Na$_v$ 1.7 and Na$_v$ 1.3) and central (Na$_v$1.7)
Stabilization of hyperexcitable neuronal membranes
Inhibition of neuronal firing
Binds CRMP-2 changes in axonal outgrowth
Advantages of Treatment with Lacosamide
Excellent oral bioavailability
Minimal serum protein binding
Excreted unchanged by the kidney
Drug-drug interaction are minimal
No regular blood tests are recommended

VGSC: Voltage-Gate Sodium Channels; CRMP-2: Collapsing Response Mediator Protein 2.

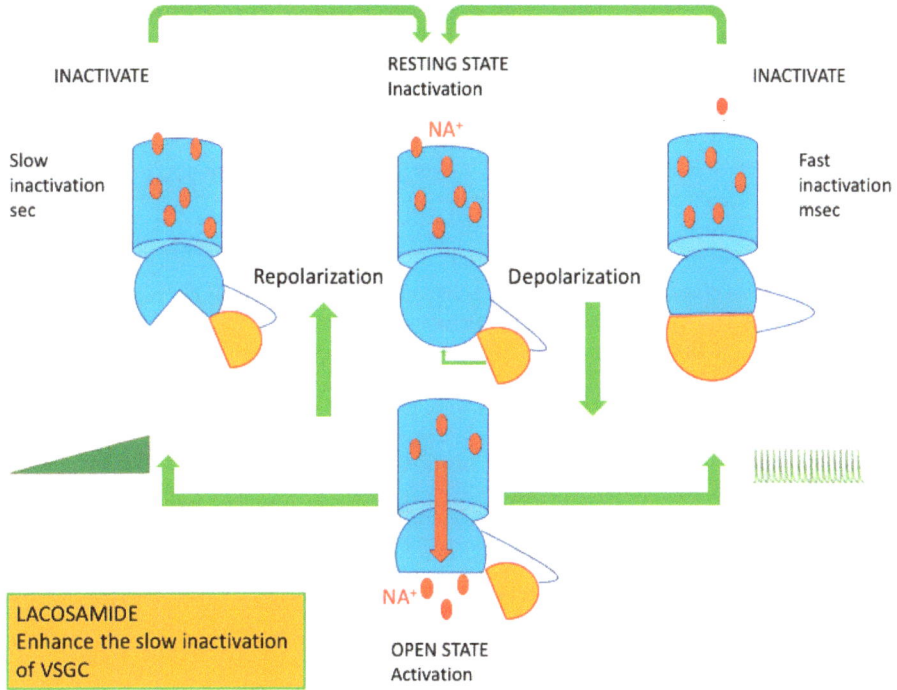

Figure 1. Lacosamide: Mechanism of action. From Dr D. Adamo et Dr. N. Coppola.

2. Presentation of the Case

A 60-year-old female patient was referred to the Oral Medicine Unit of the University Hospital of Federico II of Naples due to an acute exacerbation of TN, previously diagnosed by the Neurology Unit of our University.

The severe and unilateral pain was in the right side of her face, characterized by paroxysms of a strong stabbing nature extending in the sensory distribution of the maxillary and mandibular trigeminal right area and described as like an electric shock. The pain lasted only for seconds, occurring over intervals of a few minutes, and was triggered by swallowing or touching the affected area, without any autonomic or other neurological symptoms.

The patient was refractory to previous treatments (gabapentin, carbamazepine, lamotrigine and pregabalin) in terms of both the absence of any pain relief and the appearance of adverse events of leukopenia. We requested a routine blood test including a Complete Blood Count (CBC), glucose, electrolyte, blood urea nitrogen and creatinine levels, the erythrocyte sedimentation rate and an ECG evaluation. All results proved to be within normal limits. A Magnetic Resonance Imaging (MRI) of the brain and brainstem with and without intravenous paramagnetic contrast was required and proved to be normal except for a few subcortical white matter hyperintensities. An oro-facial evaluation and battery scales for an assessment of pain intensity and an evaluation of psychological profile were performed.

We started the treatment with a low dosage of oxcarbazepine (300 mg/daily) and re-evaluated the patient with a request for a blood examination after one month. We discovered that the WBC count had decreased from 5500 to 1900 cells/µL. We waited until the WBC had returned to a normal level and we started the treatment with 100 mg twice/daily of LCM. Pain relief was obtained in three weeks of treatment without any variations in the white blood cell count. The value of white blood cells was evaluated every month for the first six months of treatment and no changes have been detected. Currently, the patient takes a maintenance dosage of 100 mg/daily without any adverse events, remaining in a state of complete well-being without pain and with an improvement of psychological profile.

3. Conclusions

LCM has shown efficacy and a good safety profile. It may be useful for patients who are either subject to refractory TN which is not responsive to conventional therapy or are affected by significant adverse side effects.

References

1. Dieleman, J.P.; Kerklaan, J.; Huygen, F.J.; Bouma, P.A.; Sturkenboom, M.C. Incidence rates and treatment of neuropathic pain conditions in the general population. *Pain* **2008**, *137*, 681–688.
2. Obermann, M. Treatment options in trigeminal neuralgia. *Ther. Adv. Neurol. Disord.* **2010**, *3*, 107–115.
3. Rogawski, M.A.; Tofighy, A.; White, H.S.; Matagne, A.; Wolff, C. Current understanding of the mechanism of action of the antiepileptic drug lacosamide. *Epilepsy Res.* **2015**, *110*, 189–205.
4. Niespodziany, I.; Leclère, N.; Vandenplas, C.; Foerch, P.; Wolff, C. Comparative study of lacosamide and classical sodium channel blocking antiepileptic drugs on sodium channel slow inactivation. *J. Neurosci. Res.* **2013**, *91*, 436–443.

© 2019 by the authors. Licensee MDPI, Basel, Switzerland. This article is an open access article distributed under the terms and conditions of the Creative Commons Attribution (CC BY) license (http://creativecommons.org/licenses/by/4.0/).

Extended Abstract

Methotrexate-induced Plasma Cell Mucositis: A Case Report of a Previous Undescribed Correlation [†]

Alessandro d'Aiuto [1,*], Maria Pellilli [2], Marta Dani [1], Fabio Croveri [1], Andrea Boggio [1], Vittorio Maurino [1] and Lorenzo Azzi [1]

[1] Department of Medicine and Surgery, University of Insubria, ASST Sette Laghi, Dental clinic, Unit of Oral Medicine and Pathology, 21100 Varese, Italy; marta.dani92.md@gmail.com (M.D.); fabio.croveri@icloud.com (F.C.); daiuto.alessandro1992@gmail.com (A.B.); vittorio.maurino@gmail.com (V.M.); lorenzoazzi86@hotmail.com (L.A.)

[2] Department of Medicine and Surgery, University of Insubria, ASST Sette Laghi, Unit of Pathology, 21100 Varese, Italy; maria.pellilli@yahoo.it

* Correspondence: daiuto.alex@gmail.com; Tel.: +39-0332825623

[†] Presented at the XV National and III International Congress of the Italian Society of Oral Pathology and Medicine (SIPMO), Bari, Italy, 17–19 October 2019.

Published: 12 December 2019

1. Case Presentation

An 85-year-old man was referred at our Oral Medicine Unit for the presence of painful, multiple ulcerative lesions in the mouth.

The patient reported the onset of the lesions several weeks before, and he was unable to eat or drink normally, thus being hospitalized due to electrolyte imbalance.

Clinically, multiple ulcerative lesions were observed bilaterally on the buccal mucosa and also on the left soft palate (Figure 1). In addition, mild angular cheilitis-like fissures were observed on the lips. Neither the oropharyngeal and the upper aero-digestive districts nor the skin showed any other lesions.

The patient's anamnesis reported the presence of Rheumatoid arthritis, treated with methotrexate (15 mg IM weekly).

Differential diagnosis included oral blistering diseases, chronic ulcerative stomatitis, methotrexate-induced ulceration, methotrexate-induced lymphoproliferative disorder, chronic granulomatous conditions, endocrinal or nutritional alterations.

An incisional biopsy of the affected area was performed, along with DIF analysis, to confirm or rule out several autoimmune blistering disorders. The lesions healed after methotrexate was discontinued and topical corticosteroid therapy was started.

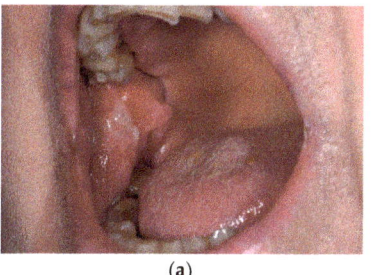 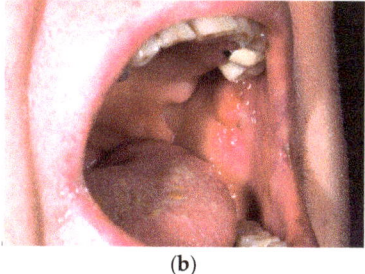

(a) (b)

Figure 1. (**a**,**b**) Clinical aspect of the lesions. Multiple ulcerations with perilesional velvety, sometimes cobblestone-looking, inflamed mucosa.

2. Histopathology

The histopathological examination revealed the presence of a chronic deep inflammatory infiltrate, completely composed of plasma cells, a feature that was confirmed by immunohistochemical analysis with CD138 (Figure 2). The infiltrate was polyclonal, with the production of both lambda and kappa chains, excluding thus any lymphoproliferative disorder or plasmacytoma.

In situ hybridization for detecting the presence of EBV infection provided negative results.

The final diagnosis was Plasma Cell Mucositis (PCM) [1]. Occasional relapses are managed with clobetasol ointment 0.05%.

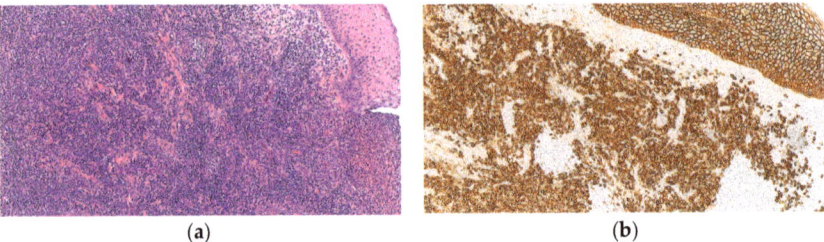

(a) (b)

Figure 2. Histopathological features of PCM. (**a**) H&E stain revealed the presence of a subepithelial inflammatory infiltrate composed completely by plasma cells. (**b**) Immunohistochemical analysis with CD138 confirmed the prevalence of the plasma cells within the inflammatory infiltrate.

3. Conclusions

PCM is a very rare immunological disorder of the upper aerodigestive tract that can affect the oral cavity [2]. This case report is the first one that describes the onset of PCM associated to Methotrexate therapy, thus different from Methotrexate-induced mucocutaneous ulcer [3].

Conflicts of Interest: The authors declare no conflict of interest.

References

1. Solomon, L.W.; Wein, R.O.; Rosenwald, I.; Laver, N. Plasma cell mucositis of the oral cavity: report of a case and review of the literature. *Oral Surg. Oral Med. Oral Pathol. Oral Radiol. Endodontol.* **2008**, *106*, 853–860, doi:10.1016/j.tripleo.2008.08.016.
2. Gasparro, R.; Adamo, D.; Masucci, M.; Sammartino, G.; Mignogna, M.D. Use of injectable platelet-rich fibrin in the treatment of plasma cell mucositis of the oral cavity refractory to corticosteroid therapy: A case report. *Dermatol. Ther.* **2019**, *32*, e13062, doi:10.1111/dth.13062.
3. Ravi, P.Y.; Sigamani, E.; Jeelani, Y.; Manipadam, M.T. Methotrexate-associated Epstein-Barr virus mucocutaneous ulcer: a case report and review of literature. *Indian J. Pathol. Microbiol.* **2018**, *61*, 255–257.

© 2019 by the authors. Licensee MDPI, Basel, Switzerland. This article is an open access article distributed under the terms and conditions of the Creative Commons Attribution (CC BY) license (http://creativecommons.org/licenses/by/4.0/).

Extended Abstract

445 nm Blue Laser in Excisional Biopsy of a Large Lipoma of the Mouth Floor †

Leonardo D'Alessandro [1,*], Francesca Graniero [1,*], Gian Marco Podda [1,*], Gaspare Palaia [1,*], Gianluca Tenore [1,*], Cira Rosaria Tiziana Di Gioia [2,*] and Umberto Romeo [1,*]

1. Department of Oral and Maxillofacial Sciences, Sapienza University of Rome, 00161 Rome, Italy
2. Department of Radiological, Oncological and Anatomopathological Sciences, Sapienza University of Rome, 00161 Rome, Italy
* Correspondence: dalessandro.1634449@gmail.com (L.D.); francesca.graniero@outlook.it (F.G.); g.m.podda@hotmail.it (G.M.P.); gaspare.palaia@uniroma1.it (G.P.); gianlucatenore@gmail.com (G.T.); cira.digioia@uniroma1.it (C.R.T.D.G.); umberto.romeo@uniroma1.it (U.R.)
† Presented at the XV National and III International Congress of the Italian Society of Oral Pathology and Medicine (SIPMO), Bari, Italy, 17–19 October 2019.

Published: 12 December 2019

Lipoma is a benign mesenchymal neoplasm, that origin from adipocytes, uncommon in oral cavity (less of 5% of the cases). It has a slow growth and often it is asymptomatic, so the diagnosis could be late. In oral cavity, buccal mucosa, tongue and mouth floor are often involved [1]. Lipoma is a submucosal lesion, sessile or pedunculated, covered by healthy mucosa, consisted of a yellow, lobulated, circumscribed and soft mass, in most cases not exceeding 1cm in diameter. Histologically, it is a mature white adipose tissue without atypia and it can be found in the variant of fibrolipoma [2]. When an oral lipoma occurs, a surgical treatment is required.

A 46-year-old female patient with a negative medical and dental history came to our observation.

The patient reported a painless swelling at mouth floor, occurring one year ago, that troubled chewing and swallowing.

At oral examination it appeared like a 3 cm × 2.5 cm sessile lesion covered by normal mucosa, from which shine through a yellow color (Figure 1).

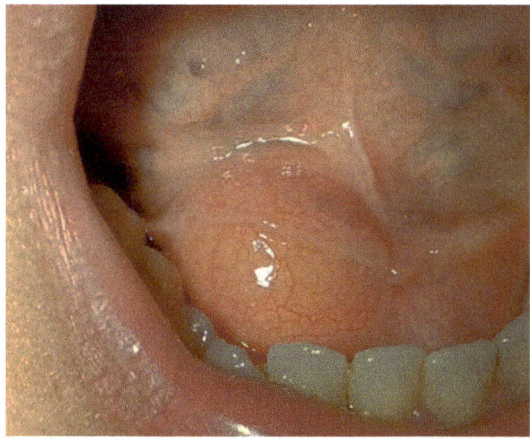

Figure 1. Intraoral view of the lesion.

At Ultrasound examination, the lesion appeared hyper-echoic and not infiltrated [3].

At Magnetic Resonance Imaging (MRI), at T1-weighted, signal was high. The lesion was placed upper than muscles plane, with major longitudinal axis of 27 mm, next to genioglossus muscle and beyond the median raphe.

An excisional biopsy was scheduled; the surgical approach was made through 445 nm diode blue laser (K-laser, Eltech s.r.l., Treviso, Italy) at 2Watt in Continuous Wave, with 320 µm optical fiber.

Blue laser has been used to make the initial vertical incision (Figure 2), it works by layers so it can isolate the superficial mucous layer from the lesion capsule, the lesion was isolated with Metzenbaum scissors and excised by laser. Laser permitted the control of bleeding, since its optimal hemostatic effect due to wavelength congenial to hemoglobin.

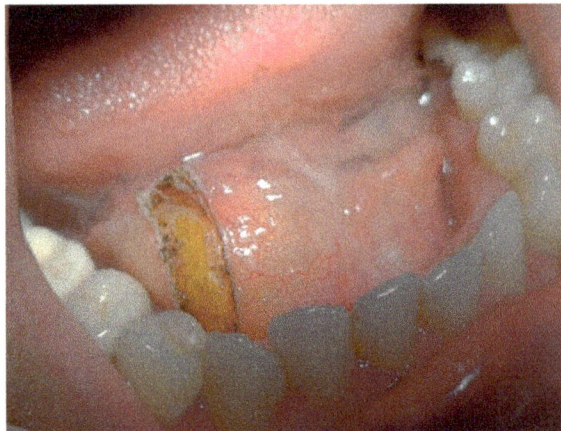

Figure 2. Laser vertical incision.

After the excision, the residual cavity was filled with iodoform gauze and 3-0 suture was applied. After 4 days, the gauze was removed in relation and an almost complete healing was observed.

The specimen was histologically analyzed and it has resulted as lipoma.

Two months follow-up shown a complete wound healing and no recurrences.

Diode blue laser, used to approach to benign neoplasms like lipomas, can offer an advantage both in the isolation of the lesion with respect to the surrounding tissues and in the intraoperative visibility thanks to the coagulation.

Conflicts of Interest: the authors declare no conflict of interest.

References

1. Manor, E. Oral lipoma: analysis of 58 new cases and review of the literature. *Ann. Diagn. Pathol.* 2011, 15, 257–261, doi:10.1016/j.anndiagpath.2011.01.003
2. Naruse, T. Lipomas of the oral cavity: clinicopathological and immunohistochemical study of 24 cases and review of the literature. *Indian J. Otolaryngol. Head Neck Surg.* **2015**, *67* (Suppl. 1), 67–73, doi:10.1007/s12070-014-0765-8.
3. Romeo, U. Color-Doppler ultrasound in the evaluation of oral lesions. *J. Ultrasound* **2017**, *20*, 351–352, doi:10.1007/s40477-017-0273-2.

 © 2019 by the authors. Licensee MDPI, Basel, Switzerland. This article is an open access article distributed under the terms and conditions of the Creative Commons Attribution (CC BY) license (http://creativecommons.org/licenses/by/4.0/).

Extended Abstract

Oral Granular Cell Tumour: A Case Report †

Marta Dani [1,*], **Maria Pellilli** [2], **Alessandro d'Aiuto** [1], **Lucia Tettamanti** [3], **Vittorio Maurino** [3] **and Lorenzo Azzi** [1]

1. Department of Medicine and Surgery, University of Insubria, ASST Sette Laghi, Dental Clinic, Unit of Oral Medicine and Pathology, 21100 Varese, Italy; daiuto.alex@gmail.com (A.A.); lorenzoazzi86@hotmail.com (L.A.)
2. Department of Medicine and Surgery, University of Insubria, ASST Sette Laghi, Unit of Pathology, 21100 Varese, Italy; maria.pellilli@yahoo.it
3. Department of Medicine and Surgery, University of Insubria, ASST Sette Laghi, Dental Clinic, Unit of Pediatric Dentistry, 21100 Varese, Italy; lucia.tettamanti@uninsubria.it (L.T.); vittorio.maurino@gmail.com (V.M.)
* Correspondence: marta.dani92.md@gmail.com; Tel.: +39-33-282-5623
† Presented at the XV National and III International Congress of the Italian Society of Oral Pathology and Medicine (SIPMO), Bari, Italy, 17–19 October 2019.

Published: 12 December 2019

1. Case Presentation

A 16-year-old girl presented at our attention with an asymptomatic nodular lesion on the tongue. Clinically, the lesion appeared as a submucosal smooth nodule, 5 mm in diameter, covered by a pink-yellowish looking mucosa (Figure 1). On palpation, it was well defined, but not encapsulated. The lesion was located on the right posterior dorsum of the tongue, it had been slowly growing for a few months, and it was completely asymptomatic. The patient's anamnesis revealed atopic dermatitis and pityriasis rosea.

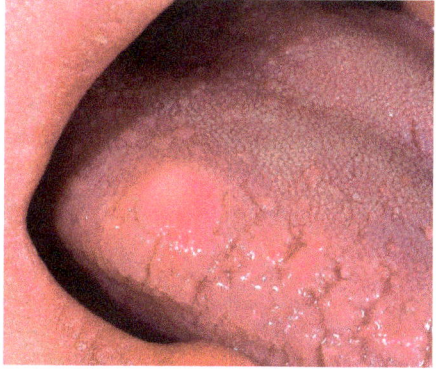

Figure 1. Clinical aspect of the lesion. A pink-yellowish submucosal nodule was detected on the posterior lateral aspect of the tongue.

Clinically, the lesion was placed in differential diagnosis with other "soft tissue tumors". Therefore, an excisional biopsy of the lesion was performed with subsequent histopathological analysis.

2. Histopathology

The connective tissue displayed a well-defined proliferation of cells with a characteristic granular cytoplasm. The cells had a characteristic granular eosinophilic cytoplasm positive to PAS staining. The neoplastic cells had a polygonal shape, large dimensions and indefinite margins

(Figure 2). The nucleus was roundish and with free chromatin. There was no nuclear atypia and the Ki-67 proliferation index was low. A peculiar histological element of the granular cell tumor was the presence of pseudoepitheliomatous hyperplasia (PEH).

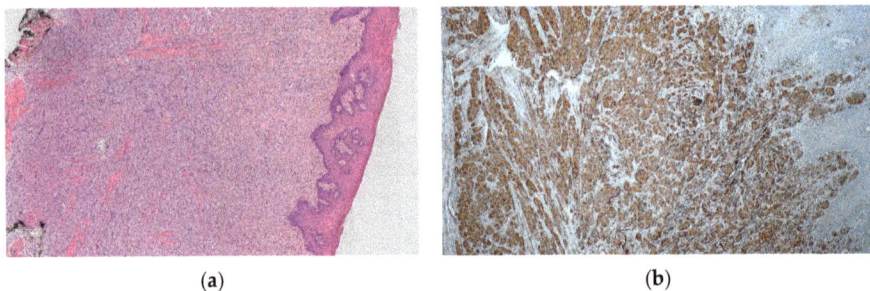

Figure 2. (a) Histopathological features of OGCT. The tumour was composed of large polygonal granular cells arranged in a pseudosyncitial pattern. The aspect is typical for GCT. (b) Immunohistochemical analysis revaled the tumour cells be positive for S-100, a marker that confirmed the origin from perineural tissues, like the Schwann cells [1].

3. Conclusions

Granular cell tumor (GCT) is a benign soft tissue neoplasm. GCT can develop throughout the body but it shows a clear preference for the head-neck district. In particular, in 50% of cases it develops on the tongue, precisely on the lateral margins and back of the tongue [2].

PEH places this lesion in a differential diagnosis with a well-differentiated squamous cell tumor [3]. In order to avoid dangerous diagnostic errors, which can lead to unnecessary lingual mutilation, it is essential to perform a correct biopsy test. Bioptic sampling must include both the epithelial lamina and the lamina propria, where the granular cell tumor develops.

Conflicts of Interest: The authors declare no conflict of interest.

References

1. Musha, A.; Ogawa, M.; Yokoo, S. Granular cell tumors of the tongue: fibroma or schwannoma. *Head Face Med.* **2018**, *14*, 1. doi:10.1186/s13005-017-0158-9.
2. Alotaibi, O.; Al Sheddi, M. Neurogenic tumors and tumor-like lesions of the oral and maxillofacial region: A clinicopathological study. *Saudi Dent. J.* **2016**, *28*, 76–79. doi:10.1016/j.sdentj.2015.12.001.
3. Al-Eryani, K.; Karasneh, J.; Sedghizadeh, P.P.; Ram, S.; Sawair, F. Lack of utility of cytokeratins in differentiating pseudocarcinomatous hyperplasia of granular cell tumors from oral squamous cell carcinoma. *Asian Pac. J. Cancer Prev.* **2016**, *17*, 1785–1787. doi:10.7314/apjcp.2016.17.4.1785.

© 2019 by the authors. Licensee MDPI, Basel, Switzerland. This article is an open access article distributed under the terms and conditions of the Creative Commons Attribution (CC BY) license (http://creativecommons.org/licenses/by/4.0/).

Extended Abstract

Particular Type of Amalgam Tattoo Associated with Rhizotomy in a Patient with Brain Malignant Tumor: A Diagnostic Dilemma [†]

Raffaella De Falco [1,]*, Luca Viganò [2], Maria Giulia Nosotti [3] and Cinzia Casu [4]

[1] RDH, Dental Hygienist, Freelancer, 44121 Ferrara, Italy
[2] RDH, Freelancer, 19121 Piacenza, Italy; mg.nosotti@gmail.com
[3] DDS, Department of Radiology, San Paolo Dental Bulding, 20122 Milan, Italy; luca.vigano1@unimi.it
[4] Private Dental Practice, 09126 Cagliari, Italy; ginzia.85@hotmail.it
* Correspondence: raffaddf67@gmail.com
† Presented at the XV National and III International Congress of the Italian Society of Oral Pathology and Medicine (SIPMO), Bari, Italy, 17–19 October 2019.

Published: 12 December 2019

1. Introduction

Oligodendrogliomas constitute 5% of all primary brain tumors and are the third most common cancer among intracranial tumors [1]. Oral mucosal melanoma is uncommon, it develops following malignant transformation of melanocytes, occasionally with a poor prognosis [2].

2. Case Report

A 66 years old female patient went to our observation for a particular pigmented lesion on the lower gum. At the anamnesis she reported suffering from temporal left oligodendroglioma, so she takes dintoina 3 times a day since 1996 when it is discovered. After surgery, in 1999 oligodentroglioma relapsed and she was undergone to chemo and radiotherapy, and relapsed again in 2003 and in 2004 she was undergone to 4 times surgical intervention and chemotherapy. In February 95, she was submitted to rhizotomy of the element 36, with reconstruction of the amalgam abutments, root canal therapy of 35 with insertion of a molten post, both elements were covered with gold-resin crowns. Seven years after she presented an evident tattoo on the vestibular adherent gingiva, less accentuated by the lingual side, corresponding to the elements 35–36.

Over the years it has had several inflammatory episodes concerning the left gingival tissue (on the same part of the brain problem) and never on the right. In January 2019, after 17 years, the pigmented lesion, stable for 16 years, was extended to element 33 e 34 both vestibular and lingual (Figures 1 and 2). Given the patient's medical history and the rapid expansion of the pigmented lesion, to exclude malignant conditions, the patient was sent to the Department of Odontostomatological Sciences of Bologna and a biopsy was performed. "Bioptic fragment of oral mucosa with deposition of blackish pigment in the chorion, mainly in the perivascular area. Lesion picture compatible with c.d. Amalgam tattoo" was the histological diagnosis.

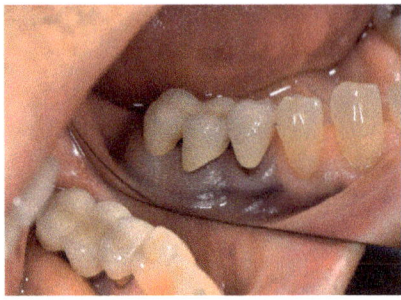

Figure 1. Pigmented lesion, vestibular side.

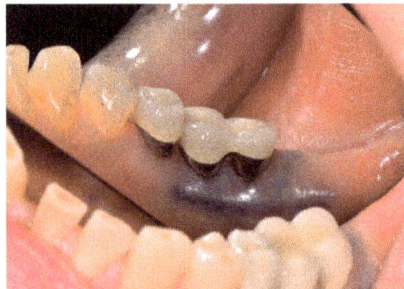

Figure 2. Pigmented lesion, lingual side.

3. Discussion

The amalgam tattoo is an iatrogenic lesion caused by the traumatic implantation of dental amalgam in soft tissues, whose diagnosis is evidenced by the position and the clinical appearance supported by the radiographic confirmation of metallic particles that are often too fine or widely dispersed to be visible on radiographs. Microscopic examination reveals that the amalgam is present in the tissues in two coexisting forms: solid metal fragments that surround the fibrous connective tissue or as numerous, thin, brown or black granules dispersed along collagen bundles around small blood vessels and nerves and are associated with a mild to moderate chronic inflammatory response, with macrophages engulfing small amalgam particles. Once the diagnosis of the amalgam tattoo has been established, no additional treatment is required except for aesthetic reasons. We have found only few cases in the scientific literature of gingival amalgam tattoo simulating malignant lesions such as melanoma [3,4].

Conflicts of Interest: The authors declare no conflict of interest.

References

1. Baran, O.; Kasimcan, O.; Oruckaptan, H. Cerebellar peduncle localized oligodendroglioma: case report and review of the literature. *World Neurosurg.* **2018**, *113*, 62–66.
2. Laimer, J.; Henn, R.; Helten, T.; Sprung, S.; Zelger, B.; Zelger, B.; Steiner, R.; Schnabl, D.; Offermanns, V.; Bruckmoser, E.; Huck, C.W. Amalgam tattoo versus melanocytic neoplasm - Differential diagnosis of dark pigmented oral mucosa lesions using infrared spectroscopy. *PLoS ONE* **2018**, *13*, e0207026.

3. Lundin, K.; Schmidt, G.; Bonde, C. Amalgam tattoo mimicking mucosal melanoma: a diagnostic dilemma revisited. *Case Rep Dent.* **2013**, *2013*, 787294.
4. Galletta, V.C.; Artico, G.; Dal Vecchio, A.M.; Lemos, C.A.; Migliari, D.A. Extensive amalgam tattoo on the alveolar-gingival mucosa. *An Bras Dermatol.* **2011**, *86*, 1019–1021.

© 2019 by the authors. Licensee MDPI, Basel, Switzerland. This article is an open access article distributed under the terms and conditions of the Creative Commons Attribution (CC BY) license (http://creativecommons.org/licenses/by/4.0/).

Extended Abstract

Effectiveness of Topical Application of Heterologous Platelet Rich Plasma (PRP) in Oral Mucous Membrane Pemphigoid. A Report of a Case [†]

Andrea Gabusi [1],*, Camilla Loi [2], Davide Bartolomeo Gissi [1], Andrea Spinelli [1], Antonio Bernardi [1] and Marina Buzzi [3]

1. Department of Biomedical and Neuromotor Sciences, Section of Oral Sciences, University of Bologna, 40125 Bologna, Italy; davide.gissi@unibo.it (D.B.G.); andrea.spinelli648@gmail.com (A.S.); Dott.abernardi@gmail.com (A.B.)
2. Department of Experimental, Diagnostic and Specialty Medicine – DIMES, Section of Dermatology, University of Bologna, 40138 Bologna, Italy; camilla.loi2@unibo.it
3. Department of Experimental, Diagnostic and Specialty Medicine – DIMES, Univeristy of Bologna, Cord Blood Bank and Cardiovascular Tissue Bank at Transfusion Medicine of S. Orsola- Malpighi Hospital; 40138 Bologna, Italy; marina.buzzi@unibo.it

* Correspondence: andrea.gabusi3@unibo.it; Tel.: +051-2088123

† Presented at the XV National and III International Congress of the Italian Society of Oral Pathology and Medicine (SIPMO), Bari, Italy, 17–19 October 2019.

Published: 12 December 2019

1. Introduction

Mucous membrane pemphigoid (MMP) is a rare, predominantly mucosal subepithelial blistering disorder triggered by autoantibody reactivity to several basement membrane antigens including BP180, BP230, laminin 332, and type VII collagen. Disease control is usually achieved with steroids and mycophenolate mofetil, azathioprine, or dapsone. However, long term use of steroids is well known to expose the patient to the risk of moderate to severe side effects.[1] For this reason, the use of alternative therapies or corticosteroid sparing agents to control disease is highly attractive[1,2]. In the present paper we describe the management of a case of mucous membrane pemphigoid treated with topical application of heterologous PRP.

2. Case Report

A 35 years old female patient affected by oral MMP and non-respondent to initial treatment with local corticosteroids was selected for heterologous PRP- based compassionate treatment with the intention to avoid systemic steroid and immunosuppressant administration. The Pemphigus Disease Area Index (PDAI) was used to assess clinical disease activity before and after treatment. Pain level changes were assessed through VAS scale from 0 to 10. Periodontal plaque index(FMPS) and bleeding on probing(FMBS) indexes were also collected to assess any improvements in periodontal situation before and after treatment. Patient underwent 7 topical applications of heterologous PRP and endpoint for clinical evaluation was set, approximately 1 month following the last session. No clinical side effects were recorded. PDAI values improved from 10 to 3. VAS score, which was reported to be 6/10 at the beginning of treatment (main complaint included the impossibility to eat crispy or hard food due to pain exacerbation and gingival bleeding) dropped to 2/10 at re-examination. Before treatment, levels of FMBS scored 30% but decreased to 9% at the end of PRP. Similarly, FMPS levels scored 26% at the beginning of treatment and dropped to 6% at the end of treatment.

3. Conclusions

Within the limits of a case report, topical applications of heterologous PRP seems to be a safe procedure for the management of patients with oral mucous membrane pemphigoid. In our case, regenerative and immunomodulatory properties of PRP achieved not only a significant improvement in both signs and symptoms of diseases but also a better control of periodontal inflammation. PRP topical application is an attractive therapeutic option to be studied at larger scale as adjunctive steroid sparing treatment for MMP.

References

1. Hussein, M.; Hassan, A.S.; Abdel Raheem, H.M.; Doss, S.S.; EL-Kaliouby, M.; Saleh, N.A.; Saleh, M.A. Platelet-rich plasma for resistant oral erosions of pemphigus vulgaris: A pilot study. *Wound Repair Regen.* **2015**, *23*, 953–955, doi:10.1111/wrr.12363.
2. Mays J.W.; Carey, B.P.; Posey, R.; Gueiros, L.A.; France, K.; Setterfield, J.; Woo, S.B.; Sollecito, T.P.; Culton, D.; Payne, A.S.; et al. World Workshop of Oral Medicine VII: A systematic review of immunobiologic therapy for oral manifestations of pemphigoid and pemphigus. *Oral Dis.* **2019**, *25* (Suppl. 1), 111–121.

© 2019 by the authors. Licensee MDPI, Basel, Switzerland. This article is an open access article distributed under the terms and conditions of the Creative Commons Attribution (CC BY) license (http://creativecommons.org/licenses/by/4.0/).

Extended Abstract

HPV-DNA Positive/p16 IHC Negative Oral Squamous Cell Carcinoma: A Case Report [†]

Di Fede Olga [1,*], Giardina Ylenia [1], Laino Luigi [3], Mascitti Marco [4], Melillo Michele [3], Capra Giuseppina [5] and Panzarella Vera [1,2]

1. Department of Surgical, Oncological and Oral Sciences (Di.Chir.On.S), University of Palermo, 90127 Palermo, Italy; ylenia.giardina@community.unipa.it (G.Y.); panzarella@odonto.unipa.it (P.V.)
2. Unit of Oral Medicine, A.O.U.P. "Paolo Giaccone", 90127 Palermo, Italy
3. Department of Clinical and Experimental Medicine, University of Foggia, 71122 Foggia, Italy; luigi.laino@unifg.it (L.L.); michele.melillo@unifg.it (M.M.)
4. Department of Clinical Specialist and Dental Sciences, Marche Polytechnic University, 60121 Ancona, Italy; m.mascitti@pm.univpm.it
5. Department PROMISE, University of Palermo, 90127 Palermo, Italy; giuseppina.capra@unipa.it
* Correspondence: odifede@odonto.unipa.it; Tel.: +0916554612-15
† Presented at the XV National and III International Congress of the Italian Society of Oral Pathology and Medicine (SIPMO), Bari, Italy, 17–19 October 2019.

Published: 12 December 2019

1. Introduction

Human Papillomavirus (HPV) is a DNA virus, belonging to papillomaviridae family. Its role in ethiopathogenesis of ano-genital cancers has been largely documented. On the contrary, HPV status in Oropharyngeal Squamous Cell Carcinoma (OPSCC) presents several controversies, supporting by a substantial heterogeneity in HPV prevalence among studies and clinical reports. These uncertainties are primarily related to two strongly interconnected reasons:

1. a non-univocal topographical classification of the OPSCC;
2. an extreme heterogeneity of the diagnostic procedures used to detect HPV status [1].

In respect to the last issue, there are several techniques to investigate HPV in human samples, with advantages and disadvantages related to clinical applicability and sensitivity/specificity in relation to HPV oncological activity. Polymerase Chain Reaction (PCR) is the most sensitivity and widely used method to detect and to genotype HPV. However, p16 immunohistochemistry (IHC) is considered a more versatile tool to indirectly measure the HPV transcriptional activity. For these reasons, its use was proposed by the 8th edition of TNM classification to stage the Head and Neck Squamous Cell Carcinomas (HNSCCs), in relation to a better prognosis of HPV positive (+ve) OPSCC recently supported in literature [2,3]. However, there are several studies reporting the low specificity of p16 IHC in relation to HPV status, potentially conditioning the appropriate treatments of patients with OPSCCs [4,5]. In this scenario, the aim of this abstract is to report a case of patient with p16 negative (-ve) OSCC, with HPV positivity by PCR investigation both in histological and on cytological samples.

2. Case Report

A Caucasian man of 61-years old, with any diseases or comorbidities noteworthy, referred to our Unit of Oral Medicine (Di.Chir.On.S./UNIPA) for the diagnostic assessment of a asymptomatic verrucous lesion localized at the right retromolar trigon (Figure 1).

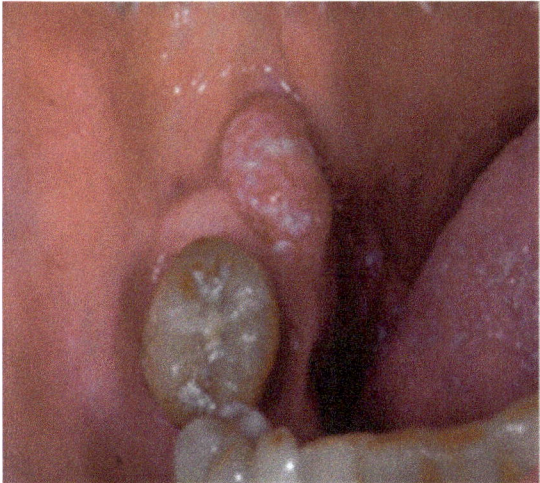

Figure 1. Verrucous lesion localized at the right retromolar trigon.

After informed consent, the patient underwent to: (i) oral rinse for HPV investigation (by PCR); and (ii) incisional biopsy both for HPV identification (by PCR and p16 IHC) and histological diagnosis. For PCR investigation, the HPV-DNA was extracted from oral rinse and biopsy sample with the use of the QIAamp Mini Kit (Qiagen). HPV genotypes were identified by the Ampliquality HPV-type express v3.0.

The histological reports was: squamous carcinoma/p16 IHC-ve.

On the contrary the microbiological examinations indicate a HPV positive status, supported by HR (High Risk) genotype 67 in both examined samples.

3. Conclusions

We report a case of confirmed diagnosis of HPV-DNA+ve/p16-ve OSCC at the right retromolar trigon. Exclusively according to the results of p16 IHC, the patient would have been resulted as negative HPV OSCC with a specific treatment. On the contrary, the use of more accurate technique of HPV investigation, highlighted a HPV-HR positivity.

This status, in addition to other therapeutic and monitoring oncological approaches potentially prospective, presupposes a specific management of the infectious condition (i.e., genital investigation, also for the partner/s) which, in the absence of the supplementary microbiological investigations conducted, would have been precluded.

In conclusion, it would be appropriate to consider, particularly for SCC in proximity to oropharyngeal areas, the necessity to use a combination of techniques (not only p16 IHC) for HPV status identification.

Conflicts of Interest: the authors declare no conflict of interest.

References

1. Lo Muzio, L.; Pelo, S. *Il Carcinoma Orale*; Grilli Editore: Foggia, Italy, 2009; pp. 58–63.
2. Huang, S.H.; O'Sullivan, B. Overview of the 8th Edition TNM Classification for Head and Neck Cancer. *Curr. Treat. Options Oncol.* **2017**, *18*, 40, doi:10.1007/s11864-017-0484-y.
3. Doescher, J.; Veit, J.A.; Hoffmann, T.K. The 8th edition of the AJCC Cancer Staging Manual: Updates in otorhinolaryngology, head and neck surgery. *HNO* **2017**, *65*, 956–961.
4. Junor, E.; Kerr, G. Benefit of chemotherapy as part of treatment for HPV-DNA-positive but p16-ve squamous cell carcinoma of the oropharynx. *Br. J. Cancer* **2012**; *106*, 358–365.

5. Qureishi, A.; Mawby, T.; Fraser, L.; Shah, K.A.; Møller, H.; Winter, S. Current and future techniques for human papilloma virus (HPV) testing in oropharyngeal squamous cell carcinoma. *Eur. Arch. Otorhinolaryngol.* **2017**, *274*, 2675–2683, doi:10.1007/s00405-017-4503-1.

© 2019 by the authors. Licensee MDPI, Basel, Switzerland. This article is an open access article distributed under the terms and conditions of the Creative Commons Attribution (CC BY) license (http://creativecommons.org/licenses/by/4.0/).

Extended Abstract
Craniofacial Fibrous Dysplasia: Diagnosis and Treatment Options †

Garibaldi Joseph, Grasso Sara *, Piazzai Matteo, Merlini Alessandro and Del Buono Caterina

U.O. Odontostomatologia, Galliera Hospital Genoa, Mura delle Cappuccine, 14, 16128 Genova, Italy; jgaribaldi@libero.it (G.J.); piazzai@libero.it (P.M.); alemerlo91@hotmail.it (M.A.); caterina.db@libero.it (D.B.C.)
* Correspondence: mini.sgrasso@gmail.com; Tel.: +39-3347010533
† Presented at the XV National and III International Congress of the Italian Society of Oral Pathology and Medicine (SIPMO), Bari, Italy, 17–19 October 2019.

Published: 12 December 2019

Fibrous dysplasia (FD) is a rare benign congenital condition characterized by the replacement of normal bone with fibrous connective tissue mixed with irregular bone trabeculae. The TAC Cone Beam represents the most effective mean for the evaluation of craniofacial FD (CFD). There are three possible approaches to the treatment of CFD: monitoring, pharmacological therapy and surgical treatment [1]. The aim of the study was to clarify on the basis of the present diagnostic means the best one to define the most appropriate therapeutic approach in the case shown in our work ie a woman of about fifty years of Brazilian origin, who came to us with an asymptomatic deformation of the left mandibular body.

In addition to the exams already in our possession, the patient was invited to perform a bone scan to evaluate radiopharmaceutical uptake and the possible presence of other skeletal lesions, as indicated by the hospital protocol for cases of suspected fibrous dysplasia. From the performed scintigraphic examination a significant asymmetry of the distribution of the osteotropic tracer on the left lower jaw was highlighted.

To confirm the diagnostic suspicion of fibrous dysplasia, a biopsy was performed.

Having ascertained the nature of the lesion, a conservative surgical approach was therefore deemed appropriate, aimed at restoring the bone size of the mandibular body without intervening on the branch, where the lesion did not have an extension such as to cause aesthetic alterations (Figure 1).

The surgery involved an osteotomy with removal of the hypertrophic segment and a curettage of the bone gap with removal of fibro-osseous particulate and of an included dental element. The flap was then closed by a single interrupted suture which was then removed after one week; the patient was monitored until the wound was completely healed [2].

The final result was satisfactory with a restoration of facial aesthetics and no neurological consequences (Figure 2).

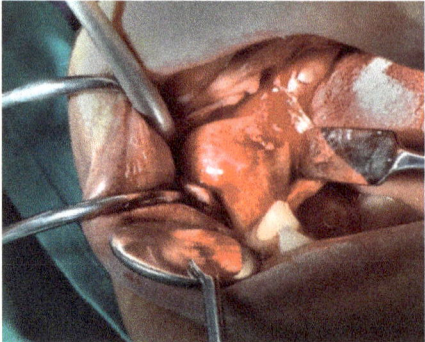

Figure 1. Clinical aspect during the surgery procedure.

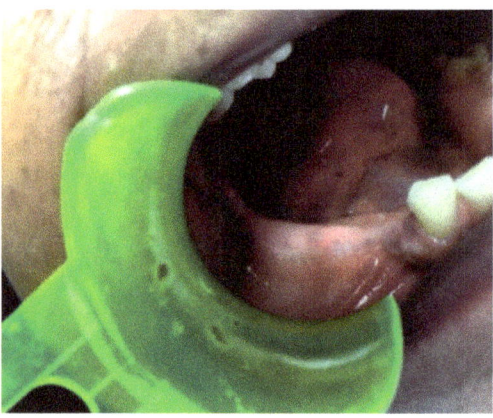

Figure 2. Final result after 6 months.

FD is a rare condition that can frequently involve the facial district. The indication for surgical treatment is not absolute, but a careful evaluation must be performed by the clinician, based on the location of the lesions and the aesthetic and functional implications that these may entail, reserving a more radical approach to cases of suspected neoplastic evolution.

Conflicts of Interest: The authors declare no conflict of interest.

References

1. Riminucci, M.; Fisher, L.W.; Shenker, A.; Spiegel, A.M.; Bianco, P.; Gehron Robey, P. Fibrous dysplasia of bone in the McCune-Albright syndrome: Abnormalities in bone formation. *Am. J. Pathol.* **1997**, *151*, 1587–600.
2. Adetayo, O.A.; Salcedo, S.E.; Borad, V.; Richards, S.S.; Workman, A.D.; Ray, A.O. Fibrous dysplasia: An overview of disease process, indications for surgical management, and a case report. *Eplasty* **2015**, *15*, e6.
3. Ricalde, P.; Magliocca, K.R.; Lee, J.S. Craniofacial fibrous dysplasia. *Oral Maxillofac. Surg. Clin. N. Am.* **2012**, *24*, 427–441.

 © 2019 by the authors. Licensee MDPI, Basel, Switzerland. This article is an open access article distributed under the terms and conditions of the Creative Commons Attribution (CC BY) license (http://creativecommons.org/licenses/by/4.0/).

Extended Abstract

Modified Double-Layered Flap Technique for Closure of an Oroantral Fistula: Surgical Procedure and Case Report [†]

Garibaldi Joseph, Grasso Sara *, Piazzai Matteo, Merlini Alessandro and Del Buono Caterina

U.O. Odontostomatologia, Galliera Hospital Genoa, Mura delle Cappuccine, 14, 16128 Genova, Italy; jgaribaldi@libero.it (G.J.); piazzai@libero.it (P.M.); alemerlo91@hotmail.it (M.A.); caterina.db@libero.it (D.B.C.)

* Correspondence: mini.sgrasso@gmail.com; Tel.: +39-3347010533

† Presented at the XV National and III International Congress of the Italian Society of Oral Pathology and Medicine (SIPMO), Bari, Italy, 17–19 October 2019.

Published: 12 December 2019

The formation of an oro-antral communication following avulsion of the lateral and posterior teeth of the maxilla is not an exceptional event in dental practice; it can undergo spontaneous resolution or the formation of a fistula that requires surgical treatment in order to create an absolutely hermetic barrier between the oral environment and the maxillary sinus [1]. The aim of the study was to provide a review of the literature on the surgical techniques currently in use for the resolution of oro-antral communications, that are the trapezoidal, rotated vestibular, rotated palatine, buccal fat pad and double-layered flap techiniques; therefore to describe the central theme of the study that is the technique of mucogingival plastic surgery called modified double-layered flap techinque moderately invasive and less subject to medium and long term recurrences [2].

We performed a review of the methods used to solve small oroantral communications (Figure 1) and carried out a case report on the alternative technique proposed by the Odontostomatology Unit of the Galliera Hospital in Genoa, ie the modified double-layered flap technique.

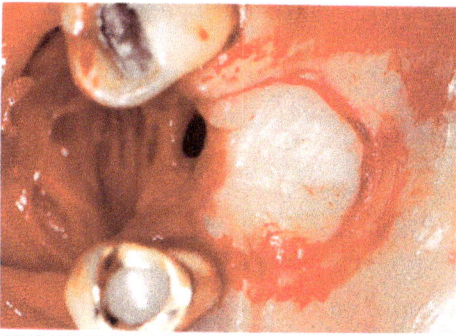

Figure 1. Example of a small oro-antral communication treated with the technique proposed.

The intent therefore remains to propose a valid protocol that is not a substitute but an alternative to the pre-existing ones, which have already been exhaustively described in the literature. The present protocol (Figure 2) has also been recognized and published by the authoritative source of the British Journal of Oral and Maxillofacial Surgery.

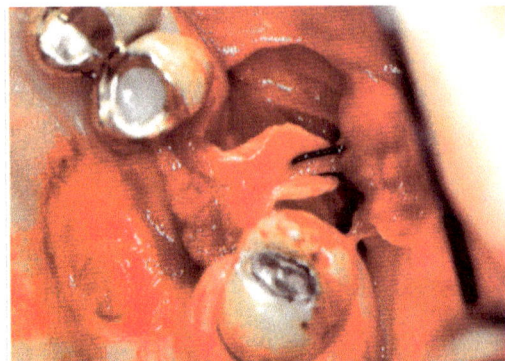

Figure 2. U-shaped anchorage to the vestibular fornix: the palatal flap constitutes the inner lining and the displacement of the vestibular flap optimally closes the primitive defect.

With regards to the satisfactory results obtained, we can state that this method is predictable.

The oro-sinus and in particular the alveolus-sinus communications are mostly sequelae of previous dental treatments and avulsions. The dentist is required to diagnose and identify the most appropriate therapeutic approach. Among the various techniques available, the modified double-layered flap technique is certainly a valid choice, as it has good predictability [3].

Conflicts of Interest: The authors declare no conflict of interest.

References

1. Abuabara, A.; Cortez, L.V.; Passeri, L.A.; de Moraes, M.; Moreira, R.W. Evaluation of different treatments for oroantral/oronasal communications: Experience of 112 cases. *Int. J. Oral Maxillofac. Surg.* **2006**, *35*, 155–158.
2. Adams, T.; Taud, D.; Rosen, M. Repair of oroantral communications by use of a combined surgical approach: Functional endoscopic surgery and buccal advancement flap/buccal fat pad graft. *J. Oral Maxillofac. Surg.* **2015**, *73*, 1452–1456.
3. Felisati, G.; Chiapasco, M.; Lozza, P.; Saibene, A.M.; Pipolo, C.; Zaniboni, M.; Biglioli, F.; Borloni, R. Sinonasal complications resulting from dental treatment: outcome-oriented proposal of classification and surgical protocol. *Am. J. Rhinol. Allergy* **2013**, *27*, e101–e106.

© 2019 by the authors. Licensee MDPI, Basel, Switzerland. This article is an open access article distributed under the terms and conditions of the Creative Commons Attribution (CC BY) license (http://creativecommons.org/licenses/by/4.0/).

Extended Abstract

The Use of Dorsum of Tongue Flap for the Closure of an Oroantral Fistula with no Contiguous Tissue Available to Be Used: Surgical Procedure and Case Report [†]

Garibaldi Joseph, Grasso Sara *, Piazzai Matteo, Merlini Alessandro and Del Buono Caterina

U.O. Odontostomatologia, Galliera Hospital Genoa, Mura delle Cappuccine 14, 16128 Genova, Italy; jgaribaldi@libero.it (G.J.); piazzai@libero.it (P.M.); alemerlo91@hotmail.it (M.A.); caterina.db@libero.it (D.B.C.)
* Correspondence: mini.sgrasso@gmail.com; Tel.: +39-3347010533
[†] Presented at the XV National and III International Congress of the Italian Society of Oral Pathology and Medicine (SIPMO), Bari, Italy, 17–19 October 2019.

Published: 12 December 2019

The formation of an oro-antral communication, a not uncommon event in dental practice, may lead to spontaneous resolution or to the formation of a fistula that requires surgical treatment in order to create an absolutely hermetic barrier between the oral environment and the maxillary sinus [1].

The aim of the study was to provide a summary review of the literature on the surgical techniques currently in use for the resolution of oro-antral communications, that are the trapezoidal, rotated vestibular, rotated palatine, buccal fat pad and double-layered flap techniques; then to describe the central theme of the study, that is the technique of mucogingival plastic surgery with the use of a dorsum of tongue flap (Figure 1) if it is not possible to use adjacent tissue to close the communication.

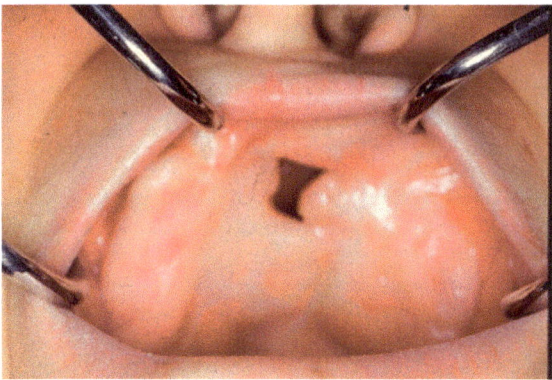

Figure 1. In this case we decided to use the technique of mucogingival plastic surgery with the use of a dorsum of tongue flap because it wasn't possible to use adjacent tissue to close the communication.

In fact, when the tissues adjacent to the oro-antral or oro-nasal communication are unsuitable for the closure of a large sized fistula, a muscolar-mucosal flap from the tongue dorsum can be used and rotated upwards.

Once this initial phase is performed, the peduncle of the flap is dissected and the excess will be repositioned to partially reconstruct the area of the dorsum of tongue used. We prepared a case report and we have photographs of the clinical situation after 20 years the surgery was performed (Figure 2).

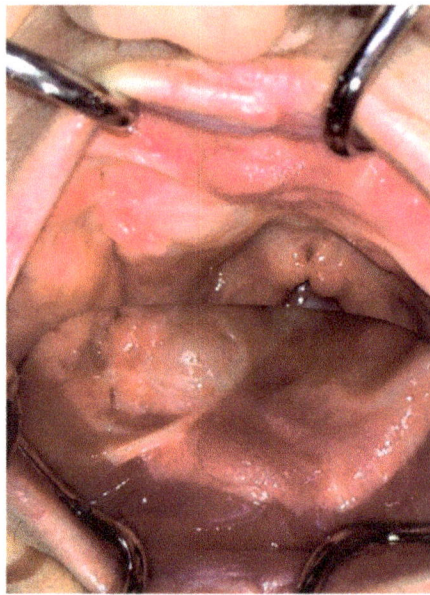

Figure 2. We see the clinical situation after 20 years the surgery was performed.

With regards to the satisfactory results obtained, we can state that this method is predictable. It's very important also the early diagnosis that allows the clinician to plan a therapeutic treatment that offers the best guarantees of success [2].

The intent therefore remains to propose a valid protocol that is not a substitute but an alternative to the pre-existing ones, already exhaustively described in literature in the specific case described [3].

Conflicts of Interest: The authors declare no conflict of interest.

References

1. Felisati, G.; Chiapasco, M.; Lozza, P.; Saibene, A.M.; Pipolo, C.; Zaniboni, M.; Biglioli, F.; Borloni, R. Sinonasal complications resulting from dental treatment: Outcome-oriented proposal of classification and surgical protocol. *Am. J. Rhinol. Allergy* **2013**, *27*, e101–e106.
2. Adams, T.; Taud, D.; Rosen, M. Repair of oroantral communications by use of a combined surgical approach: Functional endoscopic surgery and buccal advancement flap/buccal fat pad graft. *J. Oral Maxillofac. Surg.* **2015**, *73*, 1452–1456.
3. Abuabara, A.; Cortez, L.V.; Passeri, L.A.; de Moraes, M.; Moreira, R.W. Evaluation of different treatments for oroantral/oronasal communications: Experience of 112 cases. *Int. J. Oral Maxillofac. Surg.* **2006**, *35*, 155–158.

© 2019 by the authors. Licensee MDPI, Basel, Switzerland. This article is an open access article distributed under the terms and conditions of the Creative Commons Attribution (CC BY) license (http://creativecommons.org/licenses/by/4.0/).

Extended Abstract

Intraoral Salivary Gland Malignancies: Targeted Surgical Therapy Is Guided by Pre-Operative Mini-Invasive Grading [†]

Luisa Limongelli [1,*], Angela Tempesta [1], Saverio Capodiferro [1], Eugenio Maiorano [2] and Gianfranco Favia [1]

[1] Department of Interdisciplinary Medicine, University of Bari, 70124 Bari, Italy; angelatempesta1989@gmail.com (A.T.); capodiferro.saverio@gmail.com (S.C.); gianfranco.favia@uniba.it (G.F.)

[2] Department of Emergency and Organ Transplantation, University of Bari, 70124 Bari, Italy; eugenio.maiorano@uniba.it

* Correspondence: luisanna.limongelli@gmail.com; Tel.: +39-080-521-8784

[†] Presented at the XV National and III International Congress of the Italian Society of Oral Pathology and Medicine (SIPMO), Bari, Italy, 17–19 October 2019.

Published: 12 December 2019

Malignant neoplasm of the salivary glands account for 3% to 6% of all head and neck cancers. The incidence of malignant salivary gland tumors is considerably higher in minor salivary glands (MSG) accounting for 30% of all salivary cancers [1] and can present themselves in a very elusive fashion because of the heterogeneity in subsite and wide spectrum of histological subtypes. The surgical therapy depends on site, stage, and histological grading [2].

The aim of this study is to highlight the importance of achieving a preoperative histological diagnosis focusing on histological subtype and grading in order to sketch out a targeted surgical therapy.

The authors selected all cases of MSG malignancies treated from 2000 to 2018 from the Complex Operating Unit of Odontostomatology of University of Bari. All the patients in the diagnostic phase underwent to clinical examination, high definition intraoral ultrasonography, TC/MRI, pre-operative FNAB/FNAC with cytological/histological and immunohistochemical examination. Surgical therapy was: conservative for low-grade malignancies (LGM), and resective with neck dissection for high-grade malignancies (HGM).

The authors selected 146 patients. In all cases the histological diagnosis and grading was achieved in the pre-operative phase. 94 were diagnosed as LGM and 52 as HGM. About the LGM:

- Twelve cases were diagnosed as Polymorphous Low Grade Adenocarcinoma (PLGA): 9 M and 3 F, mean age 54, 10 on palate and 2 on cheek;
- Forty-one cases were diagnosed as low-grade Mucoepidermoid Carcinoma: 29 F and 12 M, mean age 37, 33 on the palate, 4 on the cheek, 2 on the lip and 2 on the tongue;
- Thirty-four cases were diagnosed as low-grade Adenoidocistic Carcinoma: 18 F and 16 M, mean age 61, 29 on the palate, 1 on the cheek, 3 on the lip and 1 on the tongue;
- Seven cases were diagnosed as Intercalated Duct Carcinoma: 3 F and 4 M, mean age 64, localized on the palate.

About the HGM:

- Nine cases were diagnosed as Clear cell Carcinoma: 4 F and 5 M, mean age 63, 8 on the palate and 1 on the lip;

- Nineteen cases were diagnosed as High-grade Mucoepidermoid Carcinoma (of which 5 clear cell high grade mucoepidermoid carcinoma, 6 on long-standing Pleomorphic Adenoma): 14 F and 5 M, mean age 41, 16 on the palate, 1 on the cheek, 1 on the lip and 1 on the tongue;
- Thirteen cases were diagnosed as high-grade Adenoidocistic Carcinoma (of which 3 clear cell high grade mucoepidermoid carcinoma, 2 on long-standing Pleomorphic Adenoma): 5 F and 8 M, mean age 54, 10 on the palate, 2 on the cheek and 1 on the lip;
- Eleven cases were diagnosed as basaloid or undifferentiated carcinoma: 5 F and 6 M, mean age 62, 9 on the palate and 2 on the cheek.

In conclusion, preoperative FNAB/FNAC for histological diagnosis of MSG malignancies is mandatory in order to make decisions on the type of surgical treatment.

Conflicts of Interest: The authors declare no conflict of interest.

References

1. Sardar, M.A.; Ganvir, S.M.; Hazarey, V.K. A demographic study of salivary gland tumors. SRM. *J. Res. Dent. Sci.* **2018**, *9*, 67–73.
2. Baddour, H.M.; Fedewa, S.A.; Chen, A.Y. Five- and 10-Year Cause-Specific Survival Rates in Carcinoma of the Minor Salivary Gland. *JAMA Otolaryngol. Head Neck Surg.* **2016**, *142*, 67–73.

© 2019 by the authors. Licensee MDPI, Basel, Switzerland. This article is an open access article distributed under the terms and conditions of the Creative Commons Attribution (CC BY) license (http://creativecommons.org/licenses/by/4.0/).

Oral Leiomyosarcoma or Low-Grade Myofibrosarcoma: Report of a Challenging Differential Diagnosis [†]

Melania Lupatelli [1], Giovanni Agrò [1], Alessandro Fornari [2] and Monica Pentenero [1,*]

[1] Department of Oncology, University of Turin, San Luigi Gonzaga Hospital, 10043 Orbassano (TO), Italy; melanialupatelli@gmail.com (M.L.); dr.giovanni.agro@gmail.com (G.A.)
[2] Pathology Division, San Luigi Gonzaga Hospital, 10043 Orbassano (TO), Italy; alefornari77@gmail.com
* Correspondence: monica.pentenero@unito.it; Tel.: +39-0119026419
[†] Presented at the XV National and III International Congress of the Italian Society of Oral Pathology and Medicine (SIPMO), Bari, Italy, 17–19 October 2019.

Published: 12 December 2019

Soft tissue spindle cell neoplasms represent less than 1% of all malignancies [1]. The pathological definitive diagnosis is challenging due to the presence of atypical histopathological and immunohistochemical (IHC) features; in more than one-third of cases the assessment of the surgical specimen allows to refine the first diagnosis from incisional biopsy.

A 31-year-old woman was referred for a painful ulcerated fast-growing swelling on the right posterior palate, with palpable cervical lymphadenopathy (Figure 1). Her past medical history is notable for thyroid and breast carcinomas. Incisional biopsy showed atypical spindle cell proliferation arranged in fascicles, with wavy nuclei and prominent nucleoli, low mitotic activity and no necrosis. IHC just demonstrated low positivity for smooth muscle actin; pan-cytokeratin, desmin, S100, CD34, CD31, CD10, ß-catenin were not found. These features were suggestive for a low-grade myofibrosarcoma (LGMS). Imaging (MRI, PET/CT, CT) showed a local spread to pterygopalatine and infratemporal fossa and regional node involvement. Neoadjuvant chemotherapy (epirubicin + ifosfamide) was performed before wide surgical resection. Assessment of the surgical specimen highlighted positive IHC for caldesmon protein leading to a final diagnosis of leiomyosarcoma (LMS). On account of multiple tumours, a NGS multi-gene panel testing including PTEN, BRCA1, BRCA2, TP53, ATM and CHEK2 was performed. ATM heterozygous mutations in exon 24 and CHECK2 homozygous mutation in exon 4 were found.

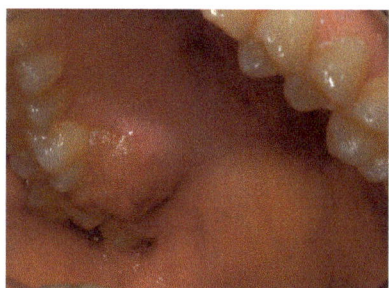

Figure 1. Submucosal ulcerated mass on the right side of the hard palate.

LMS represents approximately 3% to 10% of H&N sarcomas [2]. The common morphological pattern includes broad fascicles of spindle cells with blunt ended, brightly eosinophilic cytoplasm, cigar-shaped nuclei and occasional perinuclear vacuoles. LMS normally shows positive IHC for mesenchymal smooth muscle antigens, while desmin is variably positive. IHC for S-100 protein, epithelial, neurofilaments, factor VIII–related and angiosarcoma is usually negative. Different tumours with myofibroblastic origin and LMS share similar morphological and IHC patterns making the differential diagnosis challenging. LGMS, whose definition was revised in 2013, originates from myofibroblasts, mesenchymal spindle-shaped cells with ultrastructural features of fibroblasts and smooth-muscle cells. In LGMS, malignant myofibroblasts differ from LMS for palely eosinophilic cytoplasm and sometimes round-shaped or vescicular nuclei with indentations and small indistinct nucleoli. IHC staining shows variable positive reactions for actin and/or desmin; heavy positivity to caldesmone is more rarely observed when compared to smooth muscle cells. Wide surgical resection with clear margins is the mainstay management for oral LMS, with a five-year survival rate of 55%. Local recurrences are frequent and distant metastases occur especially in lungs. Other than this, LMS arising from the oral cavity has a high rate of regional involvement. Sometimes sarcomas occur in heritable cancer predisposition syndromes: recent studies reported multiple genes involved in the development of LMS, among which ATM and CHEK2. Genetic studies could lead to novel therapies: to date AKT/MTOR pathways seem to be the most promising potential therapeutic targets especially in high-grade LMS.

Funding: This research received no external funding.

Conflicts of Interest: The authors declare no conflict of interest.

References

1. Ko, E. Primary oral leiomyosarcoma: A systematic review and update. *J. Oral Pathol. Med.* **2019**, 1–8, doi:10.1111/jop.12858.
2. Cai, C.; Dehner, L.P.; El-Mofty, S.K. In myofibroblastic sarcomas of the head and neck, mitotic activity and necrosis define grade: a case study and literature review. *Virchows Arch.* **2013**, *463*, 827–836, doi:10.1007/s00428-013-1494-1.

© 2019 by the authors. Licensee MDPI, Basel, Switzerland. This article is an open access article distributed under the terms and conditions of the Creative Commons Attribution (CC BY) license (http://creativecommons.org/licenses/by/4.0/).

Extended Abstract

Atypical Gingival Swelling Unrelated to Plaque and Tartar: Diagnostic Difficulty and Conservative Treatment [†]

Giovanna Mosaico [1,*], Alessio Chirulli [2], Antonia Sinesi [3], Luca Viganò [4] and Cinzia Casu [5]

1. RDH, Freelancer in Brindisi, 72100 Brindisi, Italy
2. DDS, Private Dental Practice, 72013 Ceglie Messapica, Italy; info@dentistachirulli.it
3. RDH, Freelancer in Canosa di Puglia, 76012 Canosa di Puglia, Italy; antonia.sinesi@gmail.com
4. DDS, Department of Radiology, San Paolo Dental Bulding, University of Milan, 20121 Milan, Italy; luca.vigano1@unimi.it
5. DDS, Private Dental Practice, 09121 Cagliari, Italy; ginzia.85@hotmail.it
* Correspondence: gimosaico@tiscali.it
† Presented at the XV National and III International Congress of the Italian Society of Oral Pathology and Medicine (SIPMO), Bari, Italy, 17–19 October 2019.

Published: 12 December 2019

1. Introduction

In recent years, works have shown that probiotics have beneficial localized effects. Probiotic treatment is effective against diseases and infections of the skin and mucous membranes [1]. Several studies on the probiotic Lactobacillus Reuteri (L. Reuteri) have demonstrated anti-inflammatories and antimicrobials effects [2].

2. Materials and Methods

A 48-year-old female patient with non-plaque and tartar gingival papilla swelling between the upper incisors 1.1–1.2, went to our observation (Figure 1). She takes drugs for hypertension and vitamin D supplement, suffers from periodontal diseases in a good state of maintenance and was subjected to professional oral hygiene sessions every 4 months. It was decided to perform a biopsy, scheduled 2 weeks later. To restore the oral probiotic microbiota L. Reuteri have been suggested. Tablets of Lactobacillus Reuteri DSM 17938 ATCC PTA 5289 have been prescribed as home therapy twice a day, for two weeks, dissolving them slowly in the mouth after careful oral hygiene, taking care to rub the tablet with the tongue against the swollen papilla. At the check the gum appeared during the normalization phase, so the treatment was prolonged for another two weeks, and the biopsy has been postponed (Figure 2).

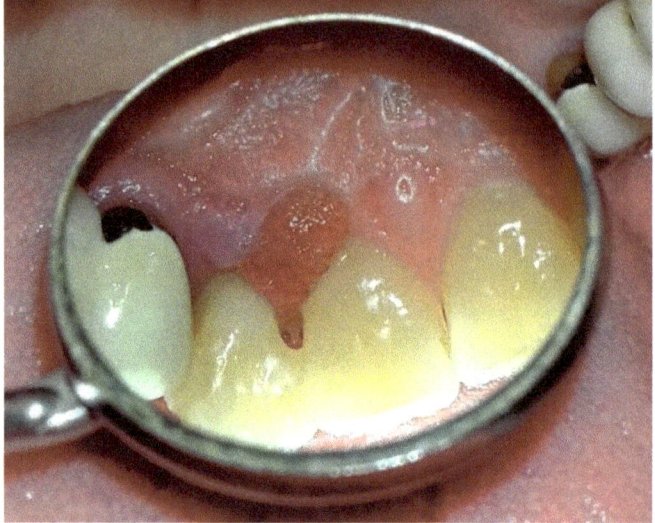

Figure 1. Patient before using probiotics L. Reuteri DSM 17938 ATCC PTA 5289.

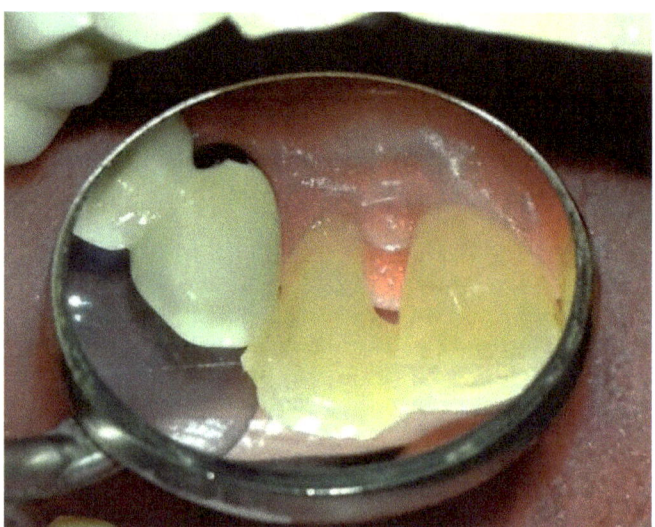

Figure 2. Patient 4 weeks after therapy with L. Reuteri DSM 17938 ATCC PTA 5289.

3. Results

4 weeks of follow-up were carried out. The gingival papilla is completely healed after treatment with L. Reuteri.

4. Discussion and Conclusions

Oral treatment with tablets containing the probiotic strain of L. Reuteri induces in most patients a significant reduction in the proinflammatory cytokine response and an improvement in clinical parameters [3]. L. Reuteri is known for its secretion of 2 bacteriocins, reuterine and reutericiclina, which inhibit the growth of a wide variety of pathogens; it also has a strong ability to adhere to host tissues with localized anti-inflammatory and antimicrobial effects [2]. Probiotics present a new ray of hope in periodontal therapy with a proven track record of safety and efficacy [2,3]. The use of topical probiotics in the treatment of gingival hypertrophy could be considered a valid alternative to conventional treatments. Further studies must be performed to confirm this starting result.

Conflicts of Interest: The authors declare no conflict of interest.

References

1. Friedrich, A.D.; Paz, M.L.; Leoni, J.; González Maglio, D.H. Message in a Bottle: Dialog between Intestine and Skin Modulated by Probiotics. *Int. J. Mol. Sci.* **2017**, *18*, 1067, doi:10.3390/ijms18061067.
2. Laleman, I.; Pauwels, M.; Quirynen, M.; Teughels, W. A dual strain Lactobacilli reuteri probiotic improves the treatment of residual pockets: a randomized controlled-clinical trial. *J. Clin. Periodontol.* **2019**, doi:10.1111/jcpe.13198.
3. Penala, S.; Kalakonda, B.; Pathakota, K.R.; Jayakumar, A.; Koppolu, P.; Lakshmi, B.V.; Pandey, R.; Mishra, A. Efficacy of local use of probiotics as an adjunct to scaling and root planing in chronic periodontitis and halitosis: A randomized controlled trial. *J. Res. Pharm. Pract.* **2016**, *5*, 86–93, doi:10.4103/2279-042X.179568.

© 2019 by the authors. Licensee MDPI, Basel, Switzerland. This article is an open access article distributed under the terms and conditions of the Creative Commons Attribution (CC BY) license (http://creativecommons.org/licenses/by/4.0/).

Extended Abstract

Labial Lesion with Heterogeneous Aspects in a Patient with Chronic Renal Failure: Diagnostic Difficulties and Literature Review [†]

Martina Salvatorina Murgia [1], Germano Orrù [2], Luca Viganò [3], Valentino Garau [1] and Cinzia Casu [4,*]

1. Department of Surgical Sciences, Session of Pedodontics and Interceptive Orthodontics, University of Cagliari, 09121 Cagliari, Italy; martina.murgia.s@gmail.com (M.S.M.); garauv@medicina.unica.it (V.G.)
2. Department of Surgical Sciences, Oral Biotechnology Laboratory, University of Cagliari, 09121 Cagliari, Italy; gerorru@gmail.com
3. Department of Radiology, University of Milan, 20121 Milano, Italy; luca.vigano1@unimi.it
4. DDS, Private Dental Practice, 09126 Cagliari, Italy
* Correspondence: ginzia.85@hotmail.it; Tel.: +39-340-842-2435
† Presented at the XV National and III International Congress of the Italian Society of Oral Pathology and Medicine (SIPMO), Bari, Italy, 17–19 October 2019.

Published: 12 December 2019

1. Introduction

Lips are a complex anatomical structure with a significant functional and aesthetic role which can be affect by numerous pathologies with different etiology [1]. Furthermore, lips often reveal to be the first site of manifestation of important systemic diseases [2]. This emphasizes the role of the dentist in the early detection of these disorders.

2. Purpose

Aims of this work are to describe an unusual case of adverse labial solar reaction and to make a brief review of the literature of pathologies included in the differential diagnosis.

3. Case Report

A 78-year-old woman patient was visited, in February 2019, for the sudden appearance of a lesion on the lower lip. Her medical history showed: chronic renal failure (CRF) and consequent cardiovascular disease. The patient underwent hemodialysis three times a week and her medications consisted of: furosemide, calcium carbonate, calcitriol and sodium polystyrene sulfonate for the CRF treatment and amlodipine, carvedilol, telmisartan and low molecular weight heparin for the cardiovascular disease treatment. Upon clinical examination, an irregular lesion was found localized on the entire surface of the lower lip characterized by active bullous lesions and completely eroded areas, partially ulcerated covered by clots and fibrin (Figure 1a). Further evaluation did not reveal other lesions in the upper lip, buccal mucosa, perioral and general skin. The patient reported a sun exposure shortly before the lesion appeared. Clinical features initially suggested a malignant or premalignant disease for which an incisional biopsy was planned that was no longer performed following the brief healing of the lesion occurred in five days (Figure 1b). The patient reported having used, as an emergency treatment, the mucilaginous gel from the parenchymatous leaves of *Aloe vera*, with subsequent immediate relief.

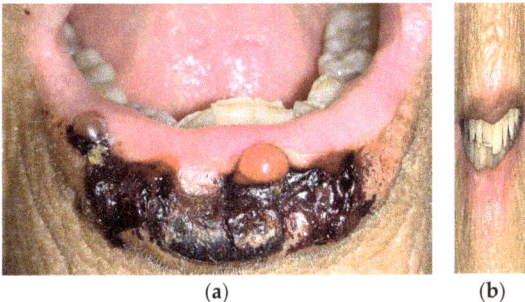

Figure 1. Clinical presentation of the lesion and subsequent healing. (**a**) Vesiculobullous lesion on the lower lip; (**b**) Resolution of the lesion.

4. Literature Review

A search of the National Library of Medicine's PubMed Database was conducted using the word "lip" together with one or more of the following key words: "cancer", "actinic cheilitis", "erythema multiforme", "burn", "bullous infection", "pemphigoid" and "pemphigus" in the last ten years. Inclusion criteria were any labial disease included in the differential diagnosis: malignant and premalignant pathologies, vesiculobullous, ulcerative and infectious diseases, adverse drug reactions and chemical burns. The initial search strategy yielded 2197 potentially relevant publications, of these, 636 studies were included in the review.

5. Result and Conclusions

The review did not lead to any clinical case similar to the one examined, which presents a very complex diagnostic decision-making approach. In literature, however, numerous cases of crusted vesiculobullous photodermatoses have been attributed to cases of pseudoporphyria (PP) [3]. Development of PP has been associated with sun exposure, CRF, hemodialysis and medications including diuretics and unlike porphrya cutanea tarda, hypertrichosis and sclerodermoid changes are not seen in PP, all factors present in this case.

Conflicts of Interest: The authors declare no conflict of interest.

References

1. Greenberg, S.A.; Schlosser, B.J.; Mirowski, G.W. Diseases of the lips. *Clin. Dermatol.* **2017**, *35*, e1–e14.
2. Vidović, V.; Nikolić, I.; Vukojević, J.; Samardžija, G.; Kukić, B.; Bogdanović, B.; Petrović, N. Unusual metastasis of esophageal cancer. *Vojnosanit. Pregl.* **2014**, *71*, 975–977.
3. Markova, A.; Lester, J.; Wang, J.; Robinson-Bostom, L. Diagnosis of Common Dermopathies in Dialysis Patients: A Review and Update. *Semin. Dial.* **2012**, *25*, 408–418.

© 2019 by the authors. Licensee MDPI, Basel, Switzerland. This article is an open access article distributed under the terms and conditions of the Creative Commons Attribution (CC BY) license (http://creativecommons.org/licenses/by/4.0/).

Extended Abstract

Sublingual Lymphangioma Mimicking a Ranula: A Case Report [†]

Marco Nisi *, Rossana Izzetti, Lisa Lardani, Lucia Scarpata, Maria Rita Giuca and Mario Gabriele

Unit of Dentistry and Oral Surgery, Department of Surgical, Medical and Molecular Pathology and Critical Care Medicine, University of Pisa, 56126 Pisa, Italy; ross.izzetti@gmail.com (R.I.); lisalardani@gmail.com (L.L.); lucia.scarpata@gmail.com (L.S.); mariarita.giuca@unipi.it (M.R.G.); mario.gabriele@med.unipi.it (M.G.)
* Correspondence: marco.nisi@unipi.it; Tel.: +39-347-354-7794
† Presented at the XV National and III International Congress of the Italian Society of Oral Pathology and Medicine (SIPMO), Bari, Italy, 17–19 October 2019.

Published: 12 December 2019

1. Background

Lymphangiomas are benign congenital malformations arising from the lymphatic system. A hamartomatous nature has been claimed for this heterogeneous cluster of disease, which is characterized by abnormal development of lymph nodes, leading to lymph collection and subsequent swelling [1,2]. In particular, in the head and neck district lymphangiomas occur with high frequency, representing almost 75% of cases. Orofacial lymphangiomas can be congenital, or develop by two years of age. The submandibular space and the posterior triangle of the neck are the most affected regions, followed by parotid area, tongue, and floor of the mouth [3]. The lesions are generally characterized by slow growth, although rapid enlargement can be observed in cases of infection or trauma. As a peculiar characteristic, lymphangiomas do not tend to spontaneous regression, often making necessary a surgical approach to the lesion, due to an impairment in function and aesthetics.

2. Case Report

A 11-year-old female patient was referred to the Unit of Dentistry and Oral Surgery, University of Pisa, for the development of a moderate swelling of the mouth floor. At clinical examination, the patient showed an asymptomatic tumefaction involving the right side of the mouth floor, causing lingual displacement (Figure 1). Considering medical history, the patient was affected by juvenile idiopathic arthritis, and was under pharmacological treatment with methotrexate and adalimumab. Several pathologic conditions entered differential diagnosis. However, sublingual ranula was strongly suspected due to the localization and the rapid development of the lesion [4].

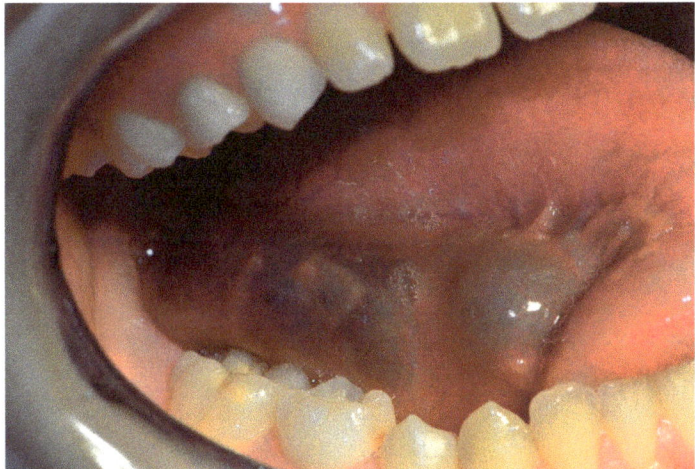

Figure 1. Clinical aspect of the lesion. A bluish fluctuant swelling covered can be observed in the right side of the mouth floor.

Head and neck MR was performed to better investigate the lesion, revealing the presence of a large mass occupying the right half of the mouth floor, and characterized by hyperintensity in T2 (Figure 2).

Initially, surgical approach was performed, with the marsupialization of the lesion and bioptic sampling. Histology revealed the presence of lymphatic channels in a connective tissue stroma, characterized by focally disorganized and thinned epithelium, and peripheral lymphoid aggregates.

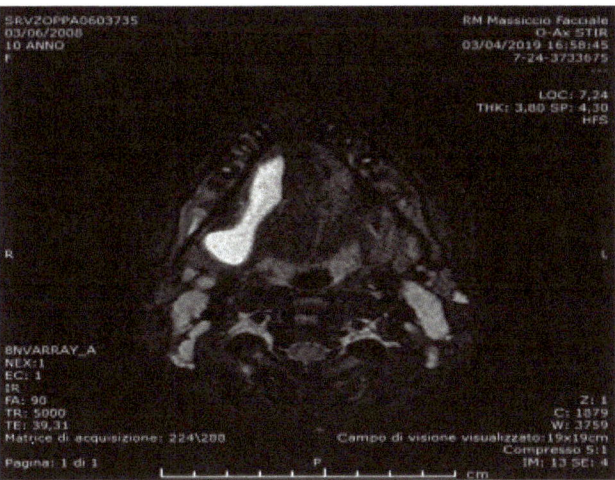

Figure 2. T2-weighted MR axial images.

Second surgery with the complete removal of the lesion was then performed. At 6-month follow-up, no signs of recurrence were observed.

3. Conclusions

Lymphangiomas usually develop early in childhood. In this peculiar case, lymphangioma developed in a 11-year-old patient, showing sudden presentation and rapid enlargement. Histology was fundamental in discriminating the nature of the lesion. Surgical approach lead to resolution of the case and complete healing at 6 months, consistently with the non-recurring nature of the disease.

Conflicts of Interest: The authors declare no conflict of interest.

References

1. Bonet-Coloma, C.; Minguez-Martínez, I.; Aloy-Prósper, A.; Rubio-Serrano, M.; Peñarrocha-Diago, M.A.; Peñarrocha-Diago, M. Clinical characteristics, treatment, and evolution in 14 cases of pediatric orofacial lymphangioma. *J. Oral Maxillofac. Surg.* **2011**, *69*, e96–e99.
2. Brennan, T.D.; Miller, A.S.; Chen, S.Y. Lymphangiomas of the oral cavity: A clinicopathologic, immunohistochemical, and electron-microscopic study. *J. Oral Maxillofac. Surg.* **1997**, *55*, 932–935.
3. Al-Abdulla, A.F.; Prabhu, S.; Al-Muharraqi, M.A.; Darwish, A.H.; Nagaraj, V. Sublingual lymphangioma that presented as a plunging ranula in a baby boy. *Br. J. Oral Maxillofac. Surg.* **2016**, *54*, 1144–1145.
4. O'Connor, R.; McGurk, M. The plunging ranula: Diagnostic difficulties and a less invasive approach to treatment. *Int. J. Oral Maxillofac. Surg.* **2013**, *42*, 1469–1474.

© 2019 by the authors. Licensee MDPI, Basel, Switzerland. This article is an open access article distributed under the terms and conditions of the Creative Commons Attribution (CC BY) license (http://creativecommons.org/licenses/by/4.0/).

Extended Abstract

Clear Cell Odontogenic Carcinoma of the Mandible: A Case Report [†]

Marco Nisi *, Rossana Izzetti and Mario Gabriele

Unit of Dentistry and Oral Surgery, Department of Surgical, Medical and Molecular Pathology and Critical Care Medicine, University of Pisa, 56123 Pisa, Italy; ross.izzetti@gmail.com (R.I.); mario.gabriele@med.unipi.it (M.G.)

* Correspondence: marco.nisi@unipi.it; Tel.: +39-3473547794
† Presented at the XV National and III International Congress of the Italian Society of Oral Pathology and Medicine (SIPMO), Bari, Italy, 17–19 October 2019.

Published: 12 December 2019

1. Background

Clear cell odontogenic carcinoma (CCOC) is a rare malignant odontogenic tumor histologically characterized by sheets and lobules of vacuolated and clear cells. To date, only 107 cases have been reported in literature since its first description by Hansen et al. in 1985 [1]. Initially classified as a benign neoplasm with the tendency to local invasion., in 2005 the World Health Organization redefined CCOC as a malignant tumor of odontogenic origin, characterized by local recurrence and presence of nodal metastases. Due to its infrequency, diagnostic criteria, protocols, and prognosis of CCOC are often not fully understood [2,3]. Additionally, CCOC shares comparable clinical and pathological characteristics with other diseases, possibly leading to misdiagnosis [4].

2. Case Presentation

A 64-year-old female patient was referred to the Unit of Dentistry and Oral Surgery, University of Pisa, for the evaluation of a gingival tumefaction in the anterior mandible. Considering medical history, the patient was affected by diabetes mellitus and hypertension, and was under pharmacologic treatment with Metformin, Levothyroxine, and Amlodipine.

At clinical examination, the patient showed an ulcerated gingival mass localized in the right mandible, associated with grade III mobility of tooth 43 (Figure 1). Panoramic radiograph showed a multilocular mixed area involving the right side of the mandible (Figure 2). CT-scan showed an osteolytic lesion with bicortical bone destruction. Incisional biopsy was performed, and on the basis of the histopathologic examination a diagnosis of ameloblastic carcinoma was made. Head and neck MR and total body PET-CT were performed to further investigate the lesion. Radical surgery included hemimandibulectomy and lymph node dissection, obtaining clear resection margins. A fibula osteoseptocutaneous flap was employed for the reconstruction of the post-surgical composite-tissue defect of the mandible. interestingly, second histology of the surgical specimen revealed a diagnosis of CCOC (positive immunodetection for anti-AE1/AE3, anti-CK19).

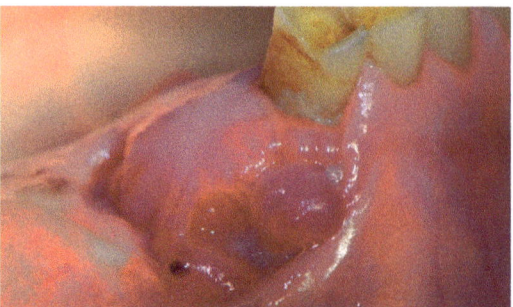

Figure 1. Clinical aspect of the lesion, involving alveolar ridge and tooth 43.

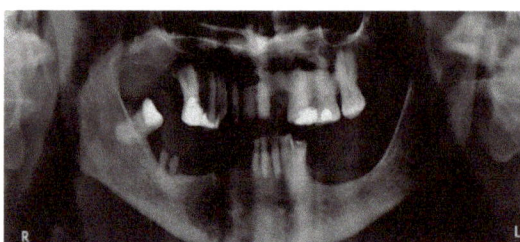

Figure 2. Panoramic radiograph. Radiolucent inhomogeneous area involving the right side of the mandible.

3. Conclusions

COCC is a rare tumor with major diagnostic difficulties. Immunohistochemistry analysis plays a key role in differential diagnosis with other odontogenic tumor. As suggested in other studies, the best treatment for COCC is wide local excision combined with regional lymph node dissection.

Conflicts of Interest: The authors declare no conflict of interest.

References

1. Guastaldi, F.P.S.; Faquin, W.C.; Gootkind, F.; Hashemi, S.; August, M.; Iafrate, A.J.; Rivera, M.N.; Kaban, L.B.; Jaquinet, A., Troulis, M.J. Clear cell odontogenic carcinoma: A rare jaw tumor. A summary of 107 reported cases. *Int. J. Oral Maxillofac. Surg.* **2019**, *48*, 1405–1410, doi:10.1016/j.ijom.2019.05.006.
2. Loyola, A.M.; Cardoso, S.V.; de Faria, P.R.; Servato, J.P.; Barbosa de Paulo, L.F.; Eisenberg, A.L.; Dias, F.L.; Gomes, C.C.; Gomez, R.S. Clear cell odontogenic carcinoma: Report of 7 new cases and systematic review of the current knowledge. *Oral Surg. Oral Med. Oral Pathol. Oral Radiol.* **2015**, *120*, 483–496, doi:10.1016/j.oooo.2015.06.005.
3. Swain, N., Dhariwal, R., Ray, J.G. Clear cell odontogenic carcinoma of maxilla: A case report and mini review. *J. Oral Maxillofac. Pathol.* **2013**, *17*, 89–94, doi:10.4103/0973-029X.110681.
4. Park, J.C.; Kim, S.W.; Baek, Y.J.; Lee, H.G.; Ryu, M.H.; Hwang, D.S.; Kim, U.K. Misdiagnosis of ameloblastoma in a patient with clear cell odontogenic carcinoma: A case report. *J. Korean Assoc. Oral Maxillofac. Surg.* **2019**, *45*, 116–120, doi:10.5125/jkaoms.2019.45.2.116.

© 2019 by the authors. Licensee MDPI, Basel, Switzerland. This article is an open access article distributed under the terms and conditions of the Creative Commons Attribution (CC BY) license (http://creativecommons.org/licenses/by/4.0/).

Extended Abstract

Photobiostimulation Therapy in Non-Responsive Oral Ulcerative Aftosis: 3 Cases Reports [†]

Maria Giulia Nosotti [1,*], Matteo Fanuli [2], Luca Viganò [3] and Cinzia Casu [4]

[1] RDH, Department of Human Sciences, Innovation and Territory, Dental Hygiene School, University of Insubria, Via Giuseppe Piatti 10, 21100 Varese, Italy
[2] RDH, Department of Biomedical, Surgical and Dental Sciences, University of Milan, 20122 Italy; matteo.fisher@hotmail.it
[3] DDS, Department of Radiology, San Paolo Dental Bulding, University of Milan, 20142 Italy; luca.vigano1@unimi.it
[4] DDS, Private Dental Practice, Cagliari, 09126 Italy; ginzia.85@hotmail.it
* Correspondence: mg.nosotti@gmail.com
† Presented at the XV National and III International Congress of the Italian Society of Oral Pathology and Medicine (SIPMO), Bari, Italy, 17–19 October 2019.

Published: 12 December 2019

1. Introduction

Ulcerative aftosis are peculiar painful lesions widely observed with common histological and pathological aspect with doubtful etiology and fewer predictable treatments [1]. These lesions are considered with primary importance in detection of some systemic issues and are often related to a wide number of pathological immune and stress-related impairments [2,3]. Elective treatment ranges from light topical anesthetics to ialuronate-based topical medications. As a bacterial thesis is proposed but never widely accepted, treatment oriented in antimicrobial way are very common, but presents few predictable results due to high heterogeneity. Here we evaluate a Photobiostimulation tissue-healing protocol [4].

2. Materials and Methods

Three patients were observed, diagnosed and treated in private practice office.

Patient history 1. Male aged 47 years, suffering from anxiety no systemic disease. He does not suffer from recurrent aphthous, referring only few episodes.

Patient history 2. Female aged 34, Hashimoto thyroiditis, recurrent aphthous stomatitis for 2 and a half years. Slow healing of mouth ulcers episodes (Figure 1).

Patient history 3. Female aged 60. Gluten intolerance, absent systemic diseases. 3 Episodes of recurrent aphtosis per year (always single and minor).

Stimulation was performed using light Laser B-cure 808 nm wavelenght, 250 mw power (Erika Carmel Ltd., Jerusalemm, Israel). It was applied directly on mouth ulcerations for 6 consecutive minutes. The device does not include projective tips, there are no probes to be applied.

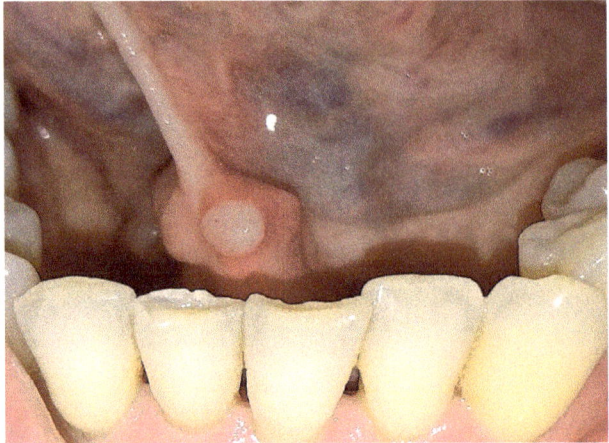

Figure 1. Patient before using the laser.

3. Results

All patients showed minor and major improvements. After 24 h, no pain symptoms were reported. After 2 days, partial or complete healing of the lesions was observed (Figure 2).

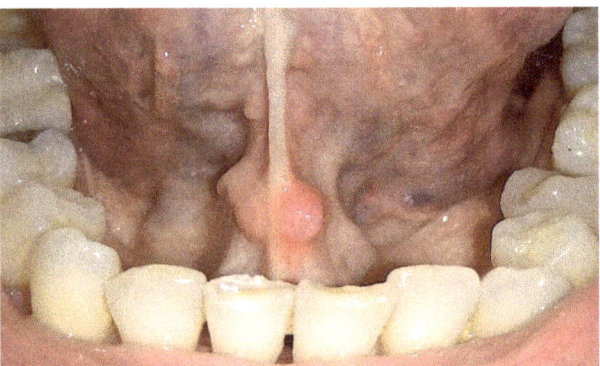

Figure 2. Patient two days after using the laser.

4. Discussion and Conclusions

Photobiostimulations protocols can be applied in the treatment of oral ulcerative lesions in recurrent cases. More reports and in vitro studies are needed to perform severe and well designed clinical protocols.

Conflicts of Interest: The authors declare no conflict of interest.

References

1. Saikaly, S.K.; Saikaly, T.S.; Saikaly, L.E. Recurrent aphthous ulceration: A review of potential causes and novel treatments. *J. Dermatol. Treat.* **2018**, *29*, 542–552.
2. Suter, V.G.A.; Sjölund, S.; Bornstein, M.M. Effect of laser on pain relief and wound healing of recurrent aphthous stomatitis: A systematic review. *Lasers Med Sci.* **2017**, *32*, 953–963.
3. Pavlić, V.; vujić-Aleksić, V.; Aoki, A.; Nežić, L. Treatment of recurrent aphthous stomatitis by laser therapy: A systematic review of the literature. *Vojnosanit. Pregl.* **2015**, *72*, 722–728.

4. Najeeb, S.; Khurshid, Z.; Zohaib, S.; Najeeb, B.; Qasim, S.B.; Zafar, M.S. Management of recurrent aphthous ulcers low-level lasers: A systematic review. *Medicina (Kaunas)* **2016**, *52*, 263–268.

 © 2019 by the authors. Licensee MDPI, Basel, Switzerland. This article is an open access article distributed under the terms and conditions of the Creative Commons Attribution (CC BY) license (http://creativecommons.org/licenses/by/4.0/).

Extended Abstract

Photodynamic Therapy in Non-Responsive Oral Angular Cheilitis: 4 Case Reports [†]

Cinzia Casu [1], Maria Giulia Nosotti [2,*], Matteo Fanuli [3] and Luca Viganò [4]

1. DDS, Private Dental Practice, 09126 Cagliari, Italy; ginzia.85@hotmail.it
2. Department of Human Sciences, Innovation and Territory, Dental Hygiene School, University of Insubria, Via Giuseppe Piatti 10, 21100 Varese, Italy
3. RDH, Department of Biomedical, Surgical and Dental Sciences, University of Milan, 20122 Milano MI, Italy; matteo.fisher@hotmail.it
4. DDS, Department of Radiology, San Paolo Dental Building, Milano, University of Milan, 20122 Milano MI, Italy; luca.vigano1@unimi.it
* Correspondence: mg.nosotti@gmail.com
† Presented at the XV National and III International Congress of the Italian Society of Oral Pathology and Medicine (SIPMO), Bari, Italy, 17–19 October 2019.

Published: 12 December 2019

1. Introduction

Detection of labial lesions in one of the more common addictional tasks for dental clinicians. These lesions are common, painful and eligible for healing trouble due to occlusion issues and comorbility factors: salivary flow rate, micotic acute opportunistics sub infections and systemic health impairments. Angular cheilitis is one of the most common labial lesion in commissural space. Painful, symmetrical and often hemorragic. Traditional treatment consist in antimicotic topical preparations such *Clotrimazole, Antimicrobial topical ointment* and *Ialuronate derived lubricants*. Advanced non responsive lesions could be treated with fillers and invasive infiltrative medication. Antimicrobial-Photodynamic therapy demonstrated high efficacy with Gram + and Gram − bacterial species. Here we suggest a clinical protocol used in well-wounded labial lesion using photosensitizers and diode light.

2. Materials and Methods

Four patients (3 females and 1 male) were treated with same protocols in private practice office. After lesion check and diagnosis of angular cheilitis labials lesions were treated with topical Toluidine Blue High Viscosity (CMS Dental, Copenaghen, Denmark) 1 mg/mL solution gently applied on the exposed dermal surface (Figure 1). After sensitizer expoxition, a 4 watt, 630 nm wave length, light emission were performed for a ten cycles of 30 s (Fotosan 630® CMS dental—Copenaghen, Denmark). After the photosensitizer was removed.

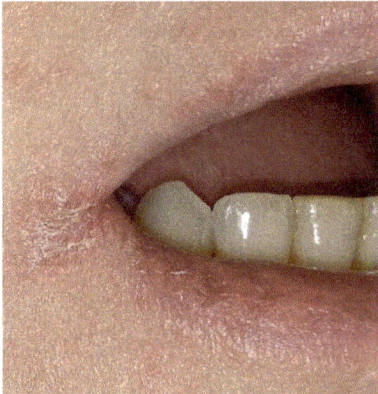

Figure 1. Patient before using Fotosan 630.

3. Results

7 days follow up were performed (Figure 2). Three patients showed complete and integral *Resitutio ad integrum*. Only one patient with severe occlusion issue shown a partial healed wound and non-integral epithelial growth after 30 days.

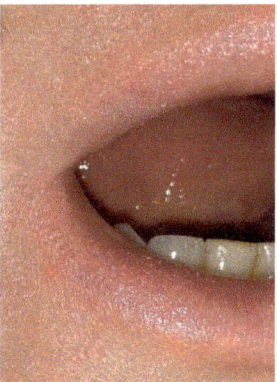

Figure 2. Patient 7 days after using Fotosan 630.

4. Discussion and Conclusions

Physical activity and chemical properties of a PDT and sensitizer adjuvancts may be used in treatment of severe non responsive labial lesions due to their antimicrobial and bacteriostatic properties.

Conflicts of Interest: The authors declare no conflict of interest.

References

1. Cabras, M.; Gambino, A.; Broccoletti, R.; Lodi, G.; Arduino, P.G. Treatment of angular cheilitis: A narrative review and authors' clinical experience. *Oral Dis.* **2019**, doi:10.1111/odi.13183.
2. Rocha, B.A.; Melo File, M.R.; Simòes, A. Antimicrobial Photodynamic Therapy to treat chemotherapy-induced oral lesions: Report of three cases. *Photodiagn. Photodyn. Ther.* **2016**, *13*, 350–352.
3. Casu, C.; Mannu, C. Atypical Afta Major Healing after Photodynamic Therapy. *Case Rep Dent.* **2017**, *2017*, 8517470.

4. Di Stasio, D.; Romano, A.; Gentile, C.; Maio, C.; Lucchese, A.; Serpico, R.; Paparella, R.; Minervini, G.; Candotto, V.; Laino, L. Systemic and topical photodynamic therapy (PDT) on oral mucosa lesions: An overview. *J. Biol. Regul. Homeost. Agents* **2018**, *32* (Suppl. 1), 123–126.

© 2019 by the authors. Licensee MDPI, Basel, Switzerland. This article is an open access article distributed under the terms and conditions of the Creative Commons Attribution (CC BY) license (http://creativecommons.org/licenses/by/4.0/).

Extended Abstract

A Case of Difficult Diagnosis: A Squamous Cell Carcinoma with Bone Exposure and Oro-sinus Communication in a Patient Receiving Alendronate [†]

Francesca Pavanelli *, Roberto Parrulli, Giuseppe Lizio, Roberta Ippolito, Salvatore Emanuele Teresi and Claudio Marchetti

Unit of Oral and Maxillofacial Surgery, Department of Biomedical and Neuromotor Science (DIBINEM), University of Bologna, 40125 Bologna, Italy; roberto.parrulli@studio.unibo.it (R.P.); giuseppe.lizio2@unibo.it (G.L.); ippolitoroberta85@gmail.com (R.I.); salvatore.teresi@studio.unibo.it (S.E.T.); claudio.marchetti@unibo.it (C.M.)
* Correspondence: Francesca.pavanelli@studio.unibo.it; Tel.: +39-3409668168
† Presented at the XV National and III International Congress of the Italian Society of Oral Pathology and Medicine (SIPMO), Bari, Italy,17–19 October 2019.

Published: 12 December 2019

Medication-Related Osteonecrosis of the Jaw (MRONJ) and Squamous Cell Carcinoma (SCC) are two distinct nosological entities that can affect the oral cavity, with different etiopathogenesis and histology. Nevertheless, the bone exposure surrounded by altered mucous tissue as a clinical manifestation of a therapy with anti-resorptive drugs can look similar to the clinical scenario observed in cases of oral carcinoma. Hence a correct diagnosis can be challenging [1,2].

A 71-year-old female patient, no smoker and no alcohol drinker, underwent treatment with Alendronate for osteoporosis for 4.5 years. The patient was referred by a Colleague with a preliminary diagnosis of MRONJ to the Oral and Maxillofacial Surgery Unit, University of Bologna, for the management of the symptomatic bone exposure. This scenario appeared about a month before and was preceded by swelling about two months before in the upper left molar alveolar region rehabilitated with a prosthetic bridge 10 years before. The clinical examination confirmed the presence of bone exposure in the reported site surrounded by an affected soft tissue on the palatal and buccal side: the mucosa appeared strongly inflamed with a granulomatous appearance and with rolled edges (Figure 1).

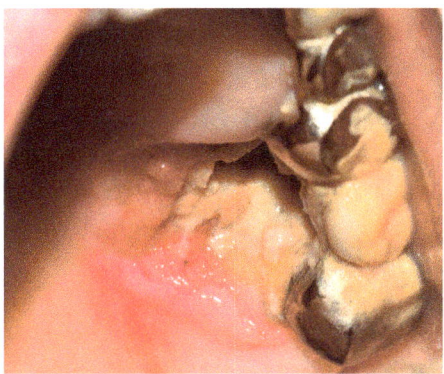

Figure 1. Clinical examination.

The panoramic x-ray did not show any sign of pathology. The Cone Beam CT revealed the presence of an oro-sinus communication of about 1 cm diameter, thickening of the sinus mucosa; the residual alveolar bone showed structural alterations with a chromatic radio-opaque and radio-lucent areas irregularly distributed (Figure 2).

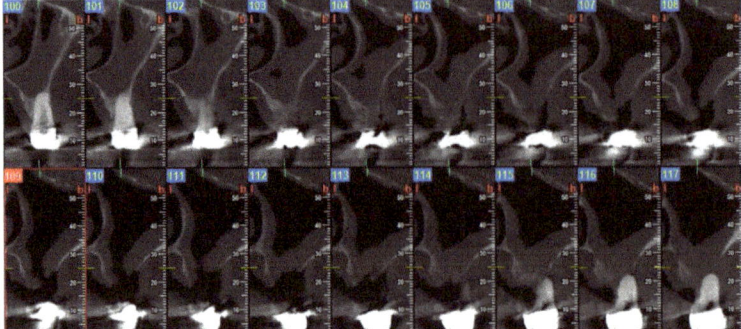

Figure 2. CBCT: paraxial cuts.

The first diagnostic option was MRONJ. Nevertheless, the particular conformation of the oral mucosa, the strong pain complained by the patient and the rapidity of occurrence usually does not match with a MRONJ scenario due to an adverse reaction to Alendronate, especially if had been assumed only for 4.5 years. The histological exam, carried out on an incisional bioptic portion of mucosa, reported the diagnosis of SCC involving the sub-mucosal layers, moderately differentiated, with signs of perineural infiltration. Hence, the patient was sent to the Maxillofacial Surgery Unit, Sant'Orsola-Malpighi Hospital of Bologna for surgical treatment in general anesthesia. We are waiting for the final histological report.

In conclusion, this singular case highlighted that the correct diagnosis is to be reached only with a careful anamnestic and clinical evaluation and a consequent peculiar diagnostic path. This goal can be reached only after a deep sharing of knowledge between the members of a specialized team that can take in consideration all the diagnostic options, even the less probable ones.

Furthermore, a question must be addressed, in our opinion, as soon as possible: which way and which relevance can the anti-resorptive drugs therapy influence the bone infiltration of an oral cavity SCC with?

Conflicts of Interest: The authors declare no conflict of interest.

References

1. Pancholi, M.; Edwards, A.; Langton, S. Bisphosphonate induced osteochemonecrosis of the jaw mimicking a tumour. *Br Dent. J.* **2007**, *203*, 87–89, doi:10.1038/bdj.2007.634
2. Tocaciu, S.; Breik, O.; Lim, B.; Angel, C.; Rutherford, N. Diagnostic dilemma between medication-related osteonecrosis and oral squamous cell carcinoma in a mandibular lytic lesion. *Br. J. Oral Maxillofac. Surg.* **2017**, *55*, e53–e57, doi:10.1016/j.bjoms.2017.08.005.

© 2019 by the authors. Licensee MDPI, Basel, Switzerland. This article is an open access article distributed under the terms and conditions of the Creative Commons Attribution (CC BY) license (http://creativecommons.org/licenses/by/4.0/).

Extended Abstract

Characterization of Bacterial Metabolites in Parotid, Submandibular/Sublingual and Whole Saliva of Healthy Subjects [†]

Margherita Eleonora Pezzi *, Rita Antonelli [1], Maria Vittoria Viani [1], Emanuela Casali [2], Thelma A. Pertinhez [2,3], Eleonora Quartieri [2], Paolo Vescovi [1] and Marco Meleti [1]

[1] Centro Universitario di Odontoiatria, Department of Oral Medicine and Surgery, University of Parma–Via Gramsci 14, 43126 Parma, Italy; rita.antonelli@hotmail.it (R.A.); mariavittoriaviani@gmail.com (M.V.V.); paolo.vescovi@unipr.it (P.V.); marco.meleti@unipr.it (M.M.)
[2] Department of Medicine and Surgery–University of Parma–Via Volturno 39, 43125 Parma, Italy; emanuela.casali@unipr.it (E.C.); thelma.pertinhez@unipr.it (T.A.P.); eleonora.quartieri@unipr.it (E.Q.)
[3] Transfusion Medicine Unit, Azienda USL–IRCCS di Reggio Emilia–Viale Umberto I, 50, 43123 Reggio Emilia, Italy
* Correspondence: margherita.pezzi@ gmail.com; Tel.: +393337388123
[†] Presented at the XV National and III International Congress of the Italian Society of Oral Pathology and Medicine (SIPMO), Bari, Italy, 17–19 October 2019.

Published: 12 December 2019

Metabolome is the comprehensive assembly of metabolites in biologic tissues and/or fluids. The study of metabolic profiles may give important information about the health status. Recently, many researches have been focused on saliva diagnostics, based in its abundant production and easier way of collection [1].

The present study reports the differences in concentration of bacterial short-chain fatty acids (SCFA) in 3 different types of saliva (parotid–PS; submandibular/sublingual–SMS; whole saliva–WS) and serum. Moreover, urea concentration, a nitrogen source, has been correlated to bacterial metabolism in PS, SMS, WS.

Ten healthy males and 10 females, aged between 20 and 25 years, were enrolled. Subjects with hyposalivation (modified Saxon's test < 5 mL/5 min) and/or with evidence of periodontal disease (periodontal screening and recording–PSR > 3 in one of the sextant) and/or caries (decay missing filled teeth–DMFT > 5 teeth with active carious lesions) were excluded. Subjects with full mouth bleeding score (FMBS) and full mouth plaque score (FMPS) > 25% were treated through non-surgical periodontal therapy, 15 days before saliva collection. Saliva and serum samples were collected between 9 and 11 a.m. Subjects were asked to avoid intense workout for 12 hours before collection, as well as food, beverage different from water and toothpaste or mouthwash. The metabolic profile was evaluated by Proton Nuclear Magnetic Resonance (^{1}H-NMR). Statistical analysis was performed through contingency tables, using the chi-squared test ($p < 0.05$ significant of association).

Among the metabolites identified in saliva, the SCFA-formate (1C), acetate (2C) and propionate (3C) were produced by bacterial flora. Acetate was the most abundant metabolite in all types of saliva followed by propionate and formate. The acetate concentration determined was 2556.4 µM in WS, 676.4 µM in PS and 354.4 µM in SMS. In serum, acetate concentration is 56.9 µM indicating the endogenous contribution of the total amount of this metabolite. Propionate was 14 times less concentrated in PS and SMS and 9 times less in WS than acetate. Formate is produced by a different metabolic pathway and were present in lower concentration and 57.7 µM in WS, 27.2 µM in PS and 22.9 µM in SMS. Differences of the 3 metabolites in the 3 typologies of saliva are statistically very significant ($p < 0.0001$).

When each SCFA is compared to urea (acetate vs. urea; propionate vs. urea and formate vs. urea) differences are statistically very significant ($p < 0.00001$), indicating an inverse relationship among the three different typologies of saliva.

To the best of our knowledge this is the first study which takes into account the metabolic contribution of SCFA to the SMS composition. Three of the metabolites most commonly produced by *bacteria* spp. are found in significantly higher concentration in WS respect to PS and SMS, indicating a differential presence of *bacteria* spp., among the three types of saliva.

Particularly, the very high concentration of acetate supports the hypothesis of the presence in the oral cavity of peculiar species of "acetogenic" *bacteria*.

Conflicts of Interest: The authors declare no conflict of interest.

References

1. Wang X. Saliva metabolomics opens door to biomarker discovery, disease diagnosis, and treatment. *Appl. Biochem. Biotechnol.* **2012**, *168*, 1718–1727, doi:10.1007/s12010-012-9891-5.

© 2019 by the authors. Licensee MDPI, Basel, Switzerland. This article is an open access article distributed under the terms and conditions of the Creative Commons Attribution (CC BY) license (http://creativecommons.org/licenses/by/4.0/).

Extended Abstract

The Outcome of Primary Root Canal Treatment in Post-Irradiated Patients: A Case Series [†]

Raffaella Castagnola [1], Irene Minciacchi [1], Cosimo Rupe [1,*], Adele Pesce [2], Maria Contaldo [3], Nicola Maria Grande [1], Luca Marigo [1] and Carlo Lajolo [1]

1. Head and Neck Department, "Fondazione Policlinico Universitario A. Gemelli–IRCCS". School of Dentistry, Università Cattolica del Sacro Cuore, 00168 Rome, Italy; castagnolaraffaella@gmail.com (R.C.); mincire@alice.it (I.M.); nmgrande@gmail.com (N.M.G.); luca.marigo@unicatt.it (L.M.); carlo.lajolo@unicatt.it (C.L.)
2. Department of Radiation Oncology, "Fondazione Policlinico A. Gemelli–IRCCS". Institute of Radiology, Università Cattolica del Sacro Cuore, Rome, 00168,, Italy; adele.pesce1987@gmail.com
3. Department of Medical-Surgical and Odontostomatological Specialties, University of Campania "Luigi Vanvitelli", Naples, 80138, Italy; maria.contaldo@gmail.com
* Correspondence: cosimorupe@gmail.com; Tel.: +393929381949
† Presented at the XV National and III International Congress of the Italian Society of Oral Pathology and Medicine (SIPMO), Bari, Italy, 17–19 October 2019.

Published: 12 December 2019

Radiotherapy (RT) is an effective treatment for head and neck cancer. A multimodal approach combining RT with surgery and chemotherapy has produced a significant increase in patients' survival rates. Patients who undergo RT may experience several adverse oral effects, among which Osteoradionecrosis (ORN) is the most severe, and tooth extractions in irradiated jaws are considered as the most severe risk factor.

It is recommended that patients undergoing head and neck RT should have a dental assessment before the start of the RT to minimize the risk of developing ORN by removing oral foci. Sadly, not all patients are referred to a dentist before the beginning of the RT; thus, some patients might require dental extractions during or after radiotherapy. In these circumstances, endodontic treatment is generally preferred to extractions even in non-restorable elements [1].

A preoperative diagnostic digital radiograph was taken to assess the apical status and the dental conditions. All periapical radiographs showed the absence of a periapical radiolucency. All therapies were performed by the same surgeon in a single visit. An intraoperative, a postoperative and follow-up radiographs were taken. The maintenance of a normal contour and width of the periodontal ligament and the absence of clinical signs and symptoms were considered radiographic and clinical success [2].

All patients, after a 277-days (584-90 days) mean follow-up, were asymptomatic, and no teeth presented periapical radiolucency. No ORN was detected in the area of treated teeth. The administered radiation dose in the periapical area of treated teeth was calculated by contouring a 0.5 cm^3 on the radiotherapy planning CT-scan. The patients received a total mean dose of 65 Gy, and the periapical mean dose was 39.36 Gy (range 22.4–63.4 Gy).

The success rate was 100%, regardless of the dose of RT received at the apex of the teeth or the time since patients underwent head and neck radiotherapy. No ORN was observed.

The novelty of this study was the measurement of the periapical radiation doses received by each treated tooth, since a correlation between apical periodontitis and radiation dose in irradiated patients has been shown [3]. However, the absence of ORN onset confirms the safety of the endodontic therapy in irradiated patients. A higher sample size is required to assess the success rate of primary root canal therapy in irradiated patients and to establish a correlation between success and radiation dose.

Conflicts of Interest: The authors declare no conflict of interest.

References

1. Lilly, J.; Cox, D; Arcuri, M.; Krell, K. An evaluation of root canal treatment in patients who have received irradiation to the mandible and maxilla. *Oral Surg. Oral Med. Oral Pathol. Oral Radiol. Endod.* **1998**, *89*, 224–226.
2. Chugal, N.; Mallya, S.; Kahler, B.; Lin, L. Endodontic Treatment Outcomes. *Dent. Clin. N. Am.* **2017**, *61*, 59–80.
3. Hommez, G.; De Meerleer, G.; De Neve, W.; De Moor, R. Effect of radiation dose on the prevalence of apical periodontitis-a dosimetric analysis. *Clin. Oral Investig.* **2012**, *16*, 1543–1547.

© 2019 by the authors. Licensee MDPI, Basel, Switzerland. This article is an open access article distributed under the terms and conditions of the Creative Commons Attribution (CC BY) license (http://creativecommons.org/licenses/by/4.0/).

Extended Abstract

A Refractory Labial Fissured Cheilitis Treated with Low Level Laser Therapy (L.L.L.T) [†]

Antonia Sinesi [1,*], Savino Cefola [2], Salvatore Grieco [3], Luca Viganò [4] and Cinzia Casu [5]

1. RDH, Freelancer in Canosa di Puglia, 76121 Canosa di Puglia, Italy
2. DDS, Private Dental Practise, 76012 Barletta, Italy; info@drsavinocefola.it
3. MDD, Dermatologist, 76012 Barletta, Italy; t.grieco@libero.it
4. RDH, Department of Biomedical, Surgical and Dental Sciences, University of Milan, 20122 Milan, Italy; luca.vigano1@unimi.it
5. DDS, Private Dental Practice, 09126 Cagliari, Italy; ginzia.85@hotmail.it
* Correspondence: antonia.sinesi@gmail.com
† Presented at the XV National and III International Congress of the Italian Society of Oral Pathology and Medicine (SIPMO), Bari, Italy, 17–19 October 2019.

Published: 12 December 2019

1. Introduction

Several studies in the literature show that Low Level Laser Therapy (L.L.L.T) is effective in the treatment of oral symptomatic soft tissue lesions such as recurrent aphthous stomatitis, herpes labialis, mucositis, erosive lichen planus.

Fissured cheilitis labialis is a lesion that affects the center of the lower lip characterized by fissures, desquamation, erythema and crusts. The aetiology is usually multifactorial, due to a primary infection and/or non-infectious causes, such as mechanical irritation, nutritional deficiency or other dermatological conditions. Drug therapy is usually a combination of topical antifungals and antibacterials and glucocorticosteroid [1].

2. Materials and Methods

A 21 years old male patient went to my observation for a refractory lesion in the lower lip. (Figure 1) The anamnesis was negative for systemic pathologies. A diagnosis of fissured cheilitis labialis was made. The patient has reported suffering for years of this pathology, which was refractory to any type of topical therapy, also on a cortisone basis. The patient was subjected to 3 sessions of L.L.L.T with Nd: YAG laser (impulses of 25 ms, 1.55 w and 50 j/cm^2) for 3 application of 4 min each whit a 3-min intermission at the day 3, 6, 10.

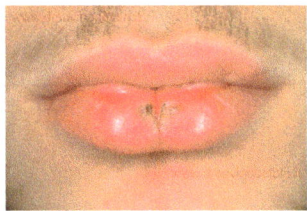

Figure 1. Labial fissured cheilitis.

3. Results

After the first photobiomodulation session we already noticed a marked improvement in the tissues. At 10 days from the first treatment with L.L.L.T the complete and total healing of the lesion was obtained with restitutio ad integrum of the affected area (Figure 2).

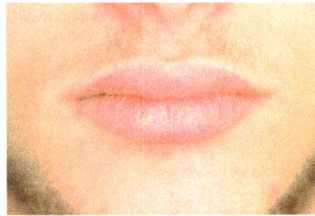

Figure 2. Labial lesion healed After L.L.L.T.

4. Conclusions

In L.L.LT, the energy of the laser beam is absorbed by the intracellular chromophore resulting in a specific response: increased cellular metabolism, improved blood circulation, increased ATP production, proliferation of all cells involved in tissue defense and repair, increased cellular respiration, increased growth factor release and new collagen synthesis. The L.L.L.T. has proved to be an extremely effective therapy in refractory fissured chelitis labialis [2–4].

Conflicts of Interest: The authors declare no conflict of interest.

References

1. Lugović-Mihić, L.; Pilipović, K.; Crnarić, I.; Šitum, M.; Duvančić, T. Differential Diagnosis of Cheilitis—How to Classify Cheilitis? *Acta Clin. Croat.* **2018**, *57*, 342–351, doi:10.20471/acc.2018.57.02.16.
2. Avci, P.; Gupta, A.; Sadasivam, M.; Vecchio, D.; Pam, Z.; Pam, N.; Hamblin, M.R. Low-level laser (light) therapy (LLLT) in skin: Stimulating, healing, restoring. *Semin. Cutan. Med. Surg.* **2013**, *32*, 41–52.
3. Hamblin, M.R. Mechanisms and Mitochondrial Redox Signaling in Photobiomodulation. *Photochem. Photobiol.* **2018**, *94*, 199–212, doi:10.1111/php.12864.
4. Spanemberg, J.C.; Figueiredo, M.A.; Cherubini, K.; Salum, F.G. Low-level Laser Therapy: A Review of Its Applications in the Management of Oral Mucosal Disorders. *Altern. Ther. Health Med.* **2016**, *22*, 24–31.

 © 2019 by the authors. Licensee MDPI, Basel, Switzerland. This article is an open access article distributed under the terms and conditions of the Creative Commons Attribution (CC BY) license (http://creativecommons.org/licenses/by/4.0/).

Extended Abstract

Nevus in the Oral Cavity [†]

Daniela Sorrentino [1,*], Sem Decani [2], Camilla Zenoni [1] and Andrea Sardella [1]

1. Dipartimento di Scienze Biomediche, Chirurgiche e Odontoiatriche, Università degli Studi di Milano, 20100 Milano, Italy; camilla.zenoni@gmail.com (C.Z.); andrea.sardella@unimi.it (A.S.)
2. ASST Santi Paolo e Carlo, Ospedale San Paolo, UO Odontostomatologia II, 20141 Milano, Italy; sem.decani@hotmail.it
* Correspondence: danielasorrentino1983@gmail.com; Tel.: +39-329-802-7264
† Presented at the XV National and III International Congress of the Italian Society of Oral Pathology and Medicine (SIPMO), Bari, Italy, 17–19 October 2019.

Published: 12 December 2019

1. Introduction

The aim of this study is to describe a clinical case of a young female patient with oral nevus. Pigmented lesions of the oral cavity represent a variety of clinical entities, ranging from physiologic changes to manifestations of systemic illnesses and malignant neoplasms. Oral pigmentations have either a melanocytic (endogenous lesions, including racial pigmentations, melanotic macules, melanocytic nevi, malignant melanoma) or a nonmelanocytic origin (exogenous lesions: amalgam tattoos). Although they may show similar clinical presentations, different treatments apply. Therefore, differential diagnoses pose a challenge in some cases. Microscopic examination can be necessary to rule out an early-stage melanoma [1–3].

2. Case

A 20-years-old woman was referred to the San Paolo Hospital in Milan due to the diagnosis of a pigmented lesion in the oral cavity during a routine examination. Intraoral examination showed the presence of a painless, exophytic neoformation, covered whit dark spots, located on left mandibular gingiva, behind the wisdom tooth (Figure 1). The consistency was elastic-soft. The patient denied smoking and alcohol consumption.

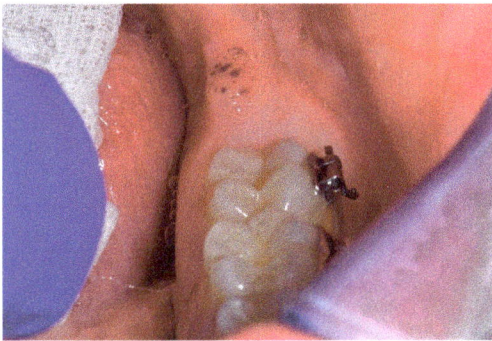

Figure 1. Clinical picture.

3. Treatment

Therefore, an excisional biopsy was performed, under local anesthesia. The incision for biopsy was elliptic, and the lesion was completely removed, and the surgical area was sutured. The removed material was fixed in 10% formalin and sent for anatomohistopathological examination. The diagnosis of composit melanocityc nevus was confirmed. After a week the suture has been removed (Figure 2).

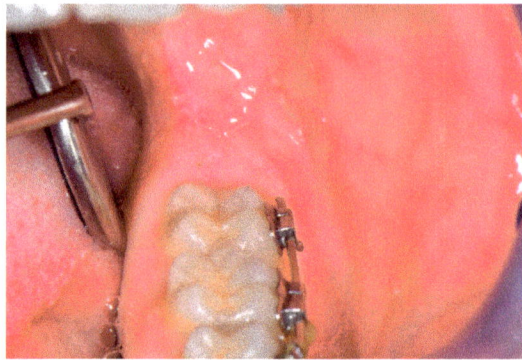

Figure 2. Postoperatory clinical picture.

4. Conclusions

Pigmented lesions represent uncommon diagnoses at an oral pathology service. Often these lesions are found in routine examination. Nevi, while uncommon, can occur in the oral cavity, and should be differentiated from other pigmented lesions, including oral melanomas that, although also rare, have a high mortality rate. When the nevi occur, they are most often found in females, in the third and fourth decade of life and the most common histological type is located on the palate and in the buccal mucosa. In this light, visual inspection of the oral cavity should be done very carefully, and the ideal treatment is the removal of the pigmented lesion whit a safety margin of 2 mm.

Conflicts of Interest: The authors declare no conflict of interest.

References

1. Popa, C.; Stelea, C.; Popa, R.; Popescu, E. Oral and perioral endogenous pigmented lesions. *Rev. Med. Chir. Soc. Med. Nat. Iasi* **2008**, *112*, 1054–1060.
2. Tavares, T.S.; Meirelles, D.P.; de Aguiar MC, F.; Caldeira, P.C. Pigmented lesions of the oral mucosa: A cross-sectional study of 458 histopathological specimens. *Oral Dis.* **2018**, *24*, 1484–1491, doi:10.1111/odi.12924.
3. Freitas, D.A.; Bonan, P.R.; Sousa, A.A.; Pereira, M.M.; Oliveira, S.M.; Jones, K.M. Intramucosal nevus in the oral cavity. *J. Contemp. Dent. Pract.* **2015**, *16*, 74–76.

 © 2019 by the authors. Licensee MDPI, Basel, Switzerland. This article is an open access article distributed under the terms and conditions of the Creative Commons Attribution (CC BY) license (http://creativecommons.org/licenses/by/4.0/).

Extended Abstract

Treatment of Symptomatic Mandibular Tori: A Case Report †

Daniela Sorrentino [1,*], **Niccolò Lombardi** [1], **Chiara Battilana** [1], **Sem Decani** [2], **Dolaji Henin** [1] **and Vincent Rossi** [1]

[1] Dipartimento di Scienze Biomediche, Chirurgiche e Odontoiatriche, Università degli Studi di Milano, 20100 Milano, Italy; niccolo.lombardi87@gmail.com (N.L.); chiara.battilana@libero.it (C.B.); dolajihenin@hotmail.it (D.H.); vincent.rossi@hotmail.it (V.R.)

[2] ASST Santi Paolo e Carlo, Ospedale San Paolo, UO Odontostomatologia II, 20100 Milano, Italy; sem.decani@hotmail.it

* Correspondence: danielasorrentino1983@gmail.com; Tel.: +32-9802-7264

† Presented at the XV National and III International Congress of the Italian Society of Oral Pathology and Medicine (SIPMO), Bari, Italy, 17–19 October 2019.

Published: 12 December 2019

1. Introduction

The aim of this paper is to describe a case of a young female patient with oral pain caused by the presence of bilateral mandibular tori. The tori are exostoses formed by a dense cortical and limited amount of bone marrow, covered with a thin and poorly vascularized mucosa. These common intraoral exostoses (prevalence 20–25%) are usually located on the palatal midline or on the lingual side of mandibular bone. In most cases this kind of exostoses don't cause pain or other symptoms and they aren't considered pathological. They can affect pronunciation and interfere with swallowing. The development of tori can be due to genetic, environmental and functional factors.

2. Case

A 34-years-old woman presented at the oral medicine department of the San Paolo Hospital in Milan. The patient reported long-lasting oral pain, exacerbated by mastication and swallowing and the previous presence of ulceration of the oral mucosa. The clinical oral examination showed a thin layer of oral mucosa covering bilateral double prominent mandibular tori, aching on palpation (Figure 1). The anterior ones measured 8 mm, the posterior one located on the left side measured 15 mm and the one on the right side measured 18 mm. Although tori removal is not always mandatory, according to the clinical features and the history of persistent pain, we decided to proceed with surgical removal.

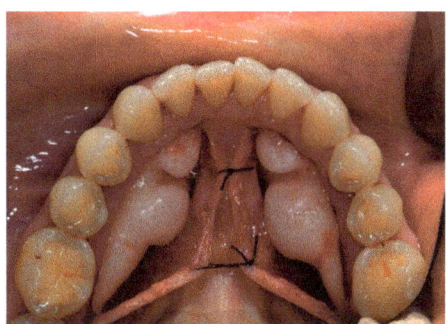

Figure 1. Mandibular tori.

3. Treatment

The surgery was conducted under local anesthesia. After a paramarginal incision, a mucoperiostal lingual flap from 19 to 30 tooth, the removal of exostosis was performed by piezoelectric surgery and sutures were placed (Figure 2). The samples were sent to pathological anatomy department for histological analysis. Postoperative medication will consist of antibiotics, analgesics and anti-inflammatory medicine.

After two weeks the patient arranged a new appointment. She referred an improvement of the symptoms. At the clinical examination soft tissue healthiness, completely healing of the surgical wound and no sign of inflammation were observed.

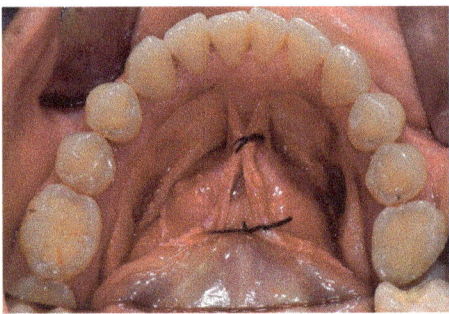

Figure 2. Post-operatory picture, after tori removal.

4. Conclusions

Although tori removal is not always necessary, there are situations in which it could be indicated: prosthetic rehabilitation or necessity of autologous bone graft. In the event that the patient reports long-lasting oral pain due to the exostosis, the removal is strongly recommended.

Conflicts of Interest: The authors declare no conflict of interest.

References

1. García-García, A.S.; Martínez-González, J.M.; Gómez-Font, R.; Soto-Rivadeneira, A.; Oviedo-Roldán, L. Current status of the torus palatinus and torus mandibularis. *Med. Oral Patol. Oral Cir. Bucal* **2010**, *15*, e353–e360.
2. Seah, Y.H. Torus palatinus and torus mandibularis: A review of the literature. *Aust. Dental J.* **1995**, *40*, 318–321.
3. Jainkittivong, A.; Langlais, R.P. Buccal and palatal exostoses: Pre-valence and concurrence with tori. *Oral Surg. Oral Med. Oral Pathol. Oral Radiol. Endodontol.* **2000**, *90*, 48–53.

 © 2019 by the authors. Licensee MDPI, Basel, Switzerland. This article is an open access article distributed under the terms and conditions of the Creative Commons Attribution (CC BY) license (http://creativecommons.org/licenses/by/4.0/).

Extended Abstract

A Case of Nivolumab-Associated Oral Lichenoid Lesions †

Andrea Spinelli *, Davide Bartolomeo Gissi, Roberto Rossi and Andrea Gabusi

Department of Biomedical and Neuromotor Sciences, Section of Oral Sciences, University of Bologna, 40125 Bologna, Italy; davide.gissi@unibo.it (D.B.G.); roberto.rossi30@studio.unibo.it (R.R.); andrea.gabusi3@unibo.it (A.G.)
* Correspondence: andrea.spinelli648@gmail.com; Tel.: +51-2088123
† Presented at the XV National and III International Congress of the Italian Society of Oral Pathology and Medicine (SIPMO), Bari, Italy, 17–19 October 2019.

Published: 12 December 2019

1. Introduction

Nivolumab is an anti-programmed cell death receptor-1 (PD-1) antibody recently approved for the treatment of melanoma, non-small-cell lung cancer, and several other solid tumours including head and neck cancers. By blocking the programmed cell death receptor-1 (PD-1), a checkpoint for the activation of the immune system, Nivolumab proved to enhance pre-existing immune responses inducing tumour recognition and attack. However, anti PD-1 blockage and related immune activation has also been related to adverse events including immunological skin, gastrointestinal, liver, endocrine system and other organ systems disorders. So far only few case reports have described lichenoid lesions triggered by nivolumab and restricted to oral cavity. In all, clinical appearance was that of multiple oral ulcers with histological features of lichenoid inflammation [1]. In this report we describe a case of reticular/erosive lichenoid lesions associated with nivolumab monotherapy.

2. Case Report

A 60 years old women was referred by her dentist for the appearance of burning oral lesions non responding to conventional treatments. At examination, reticular bilateral striae in a context of atrophic ulcerative erosion could be noted. The patient had received a diagnosis of thyroid adenocarcinoma in 2017 and had been treated with nivolumab monotherapy ever since. In 2019 she switched from a 240 mg/2 weeks regimen to a 480 mg monthly administration. 2 months after the change in the therapeutic regimen lesions appeared. The patient was also taking ranipril for blood pressure control. An incisional biopsy was taken revealing a band-like infiltrate mainly composed of lymphocytes with evidence of basal membrane aggression. Direct immunofluorescence was negative. Given the close change in cancer therapy and in absence of other putative factors, a diagnosis of lichenoid lesions associated with anti-PD-1 antibody inhibitors was made. In accordance with her oncologist, the patient preferred not to suspend nor change nivolumab therapy giving priority to tumour response. For the same reason the patient preferred to delay therapy with local/systemic steroid as symptoms did not impact significantly on her quality of life. She is currently on regular follow up.

3. Conclusions

Lichenoid lesions associated with anti-PD-1 antibody inhibitors may manifest not solely as oral ulcerative lesions but may adopt different clinical variants being thus undistinguishable from oral lichen planus. Given the recent approval of anti-programmed cell death receptor-1 (PD-1) antibodies for many solid tumors the use of anti-PD-1 antibody inhibitors is expected to grow. Consequently, despite only few cases of oral lichenoid adverse reactions have been described and yet data regarding their clinical management is still limited, dentists should be informed of their existence in order to avoid delays in diagnosis and treatment. It is still unclear if concurrent medications may have a role in the pathogenesis.

Conflicts of Interest: The authors declare no conflict of interest.

Reference

1. Obara, K.; Masuzawa, M.; Amoh, Y. Oral lichenoid reaction showing multiple ulcers associated with anti-programmed death cell receptor-1 treatment: A report of two cases and published work review. *J. Dermatol.* **2018**, *45*, 587–591, doi:10.1111/1346-8138.14205.

© 2019 by the authors. Licensee MDPI, Basel, Switzerland. This article is an open access article distributed under the terms and conditions of the Creative Commons Attribution (CC BY) license (http://creativecommons.org/licenses/by/4.0/).

Extended Abstract

Advanced Stages of Medication-Related Osteonecrosis of the Jaw: From Diagnosis to Surgical Treatment and Rehabilitation with Removable Prosthesis [†]

Angela Tempesta *, Luisa Limongelli, Saverio Capodiferro, Massimo Corsalini and Gianfranco Favia

Department of Interdisciplinary Medicine, University of Bari, 70124 Bari, Italy; luisanna.limongelli@gmail.com (L.L.); capodiferro.saverio@gmail.com (S.C.); massimo.corsalini@uniba.it (M.C.); gianfranco.favia@uniba.it (G.F.)
* Correspondence: angelatempesta1989@gmail.com; Tel.: +39-0805218784
† Presented at the XV National and III International Congress of the Italian Society of Oral Pathology and Medicine (SIPMO), Bari, Italy, 17–19 October 2019.

Published: 12 December 2019

Medication-related osteonecrosis of the Jaw (MRONJ) is an uncommon but potentially serious side effect of treatment with antiresorptive or antiangiogenic drugs among oncologic and osteoporotic patients. Clinical examination and radiological exams are mandatory for diagnosis and staging; nevertheless, even after II level exams, as Computed Tomography, MRONJ understaging is frequent [1]. Moreover, the association with pus discharge, chronic sinusitis and neurologic pathology is frequent, especially for Stage II and III lesions. To date, the treatment is still controversial, but many studies in the current literature state that only surgical removal of necrotic bone guarantees complete healing, especially for high stage lesions [1]. This surgical approach produces severe bone loss, with difficult aesthetic and functional prosthetic rehabilitation, because bone regeneration with autologous or heterologous graft cannot be performed among these patients [2]. The aim of the current study was to describe diagnostic and therapeutic protocol for patients affected by MRONJ, and subsequent rehabilitation with removable prosthesis (RP).

For the current study, authors selected 41 patients from the database of MRONJ of the Complex Operating Unit of Odontostomatology comprehending 301 patients with 389 treated lesions. The inclusion criteria were:

1. Diagnosis of Stage III MRONJ according with AAOMS criteria
2. Surgical treatment of MRONJ with complete healing after not less than 6-months follow-up
3. Total edentulism of upper and/or lower jaw after treatment

All lesions were diagnosed and treated according to our protocol. Patients' anamnestic data were collected; Rx OPT and multislice spiral CT with 3D reconstruction were performed for the staging. Patients underwent marginal bone resection including at least 1 cm of vascularized bone tissue. The depth of resection was pinpointed by the bleeding evaluation of bone tissues. Surgery was complemented by using piezosurgical device for osteoplasty and with the application of a medical device made of hyaluronic acid and amino acids. Subsequently, after complete lesions healing without recurrence, the selected patients underwent prosthetic evaluation and RP were inserted. All dentures underwent relining with soft materials to reduce risk of trauma on gingival tissue, and patients underwent periodic follow-up (once-a-month for the first year, then 3-times a year).

Overall, after a follow-up time of not less than 3 years, all patient showed complete healing of MRONJ without signs and symptoms of recurrence. Removable prosthesis allowed a good functional and aesthetic rehabilitation, with patients' satisfaction. Periodic relining with soft materials were performed (generally every 3 months). Decubitus ulcers of oral mucosa were evaluated, especially after the first insertion of prosthesis, but they always healed after dentures adjustments and relining. In these cases, application of a gel compound with hyaluronic acid and amino acids was prescribed to accelerate wound healing thus reducing the risk of MRONJ occurrence. In no one case MRONJ onset was evaluated.

Advanced Stages of MRONJ always require surgical treatment producing severe bone loss with aesthetic and functional deficit. In these cases, RP could be valid solution for patients. Nevertheless, considering that ill-fitting RP could produce MRONJ, periodic relining with soft materials and continuous adjustment are mandatory to avoid this complication.

Conflicts of Interest: The authors declare no conflict of interest.

References

1. Favia, G.; Tempesta, A.; Limongelli, L.; Crincoli, V.; Maiorano, E. Medication-related osteonecrosis of the jaw: Surgical or non-surgical treatment? *Oral Dis.* **2018**, *24*, 238–242.
2. Zirk, M.; Kreppel, M.; Buller, J.; Pristup, J.; Peters, F.; Dreiseidler, T.; Zinser, M.; Zöller, J.E. The impact of surgical intervention and antibiotics on MRONJ stage II and III—Retrospective study. *J. Craniomaxillofac. Surg.* **2017**, *45*, 1183–1189.

© 2019 by the authors. Licensee MDPI, Basel, Switzerland. This article is an open access article distributed under the terms and conditions of the Creative Commons Attribution (CC BY) license (http://creativecommons.org/licenses/by/4.0/).

Extended Abstract

MRONJ Treatment with Ultrasonic Navigation: A Case Report †

Salvatore Emanuele Teresi *, Gerardo Pellegrino, Roberto Parrulli, Agnese Ferri, Francesca Pavanelli, Riccardo Pirrotta and Claudio Marchetti

Unit of Oral and Maxillofacial Surgery, Department of Biomedical and Neuromotor Science (DIBINEM), University of Bologna, 40125 Bologna, Italy; gerardo.pellegrino2@unibo.it (G.P.); roberto.parrulli@studio.unibo.it (R.P.); agnese.ferri@studio.unibo.it (A.F.); francesca.pavanelli2@unibo.it (F.P.); riccardo.pirrotta@studio.unibo.it (R.P.); claudio.marchetti@unibo.it (C.M.)
* Correspondence: salvatore.teresi@studio.unibo.it; Tel.: +39-3209118860
† Presented at the XV National and III International Congress of the Italian Society of Oral Pathology and Medicine (SIPMO), Bari, Italy, 17–19 October 2019.

Published: 12 December 2019

Nowadays dynamic computer-based image navigation has become ordinary for hospital-based surgical specialties such as neurosurgery, otolaryngology and maxillofacial-surgery, its use for osteonecrosis curettage is considered "off-label" and isn't reported in literature. The surgical navigator is a technological-tool which relate the real anatomy of a patient to his radiological images, showing the exact three-dimensional intraoperative position of surgical-instruments. In particular, ImplaNav (BresMedical, Sydney, Australia) is composed of an infrared-camera and reference systems that are placed on the patient and on surgical-handpiece, whose position is detected in real-time by camera.

In this report we expose the innovative use of navigated ultrasonic surgery in the treatment of a bilateral medication-related osteonecrosis of the jaw (MRONJ) stage 2b [1] involving the right mandibular canal's roof in molar region, which is why the patient reported paresthesia to lower right lip, and incurred following the lower first molars' extraction in a 75-year-old male patient, subjected to 43 previous administrations of Zoledronate (from January 2014 to June 2017) for the treatment of bone metastases from stage IV follicular-thyroid carcinoma.

In order to operate with navigator's aid, the patient's pre-operative cone-beam computed-tomography (CBCT) was performed by positioning a reference system on dental arches, fixed with an impression material, according to ImplaNav protocol [2].

The surgery was conducted as in-office procedure (Figures 1 and 2) under local anesthesia, after antibiotic-prophylaxis with Amoxicillin + Clavulanic-Acid 1 g/8 h and Metronidazole 250 mg/8 h, both from 3 days before the operation. The preparation of the buccal and lingual flaps and their subsequent suturing have been performed in order to ensure an optimal vision and primary wound closure.

Compared to the traditional multi-blade burr mounted on a straight-handpiece, the well-known atraumaticity of ultrasonic-surgery allowed the respect of nerve-vascular bundle and a reduced trauma on the bone. The navigation added, to the simultaneously clinical finding of bleeding bone, the possibility of a constant comparison between clinical vision and CBCT monitor vision, allowing a rapid and complete removal of the radiographically detected altered bone.

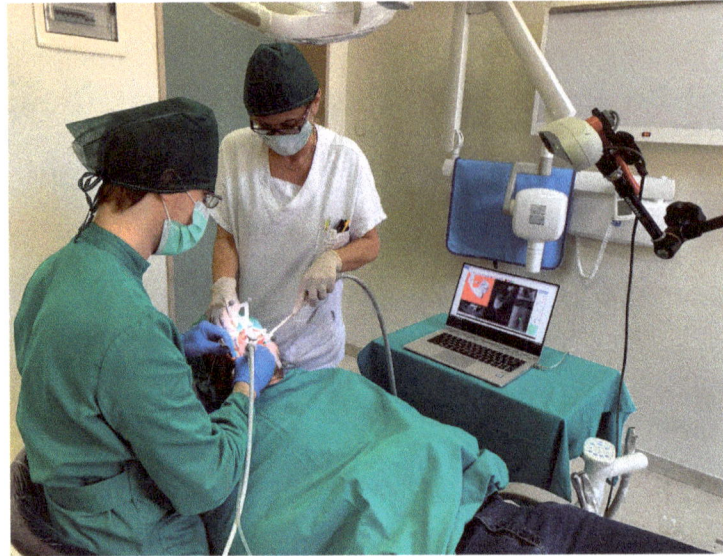

Figure 1. The reference systems placed on the patient and on the surgical-handpiece are detected by the infrared-camera. In this way, the monitor can show in real-time to the surgeon the exact three-dimensional position of the surgical-instruments compared to the preoperative-CBCT.

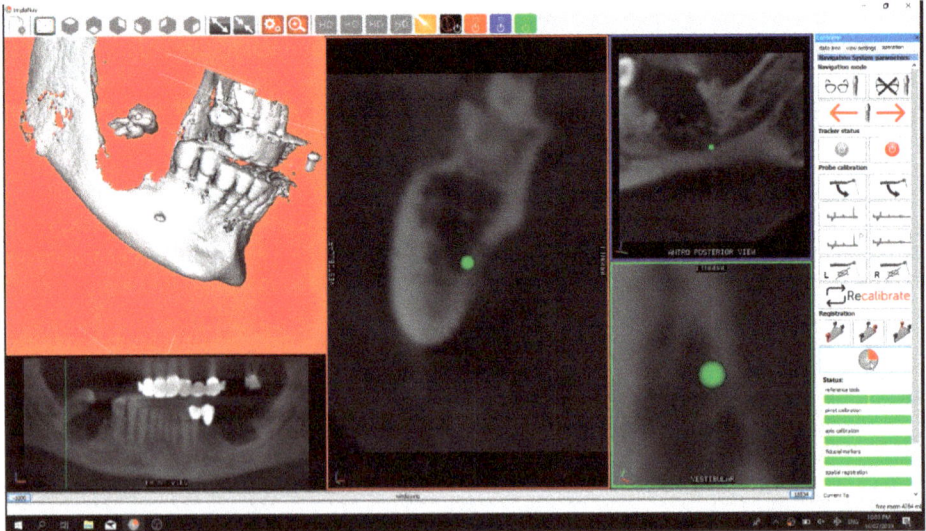

Figure 2. The intraoperative screenshot of the monitor shows the position of the ultrasonic tip compared to the preoperative-CBCT. The screenshot emphasizes how the surgeon, with the aid of the ultrasonic navigation system, can operate safely and accurately working in close proximity to the vascular-nerve bundle involved in the osteonecrosis.

The patient continued the antibiotic therapy according to aforementioned posology for another 7 days and, to date, has undergone checks at 1, 2, 4 and 12 weeks. During these visits a healing by first intention was appreciated in the absence of signs and symptoms of inflammation, however, paraesthesia remains, even if, as reported by the patient, in improvement.

The navigated ultrasonic surgery, respect to conventional free-hand surgery, has reduced the timing of surgery and tissue injury and increased the accuracy, mini-invasiveness and safety, maximizing the control of surgical-instruments and respecting the noble structures. The positive outcome of this first case of MRONJ, managed with navigated ultrasonic surgery, suggests the possibility of using this method in further cases in order to confirm the aforementioned advantages and standardize the technique.

Conflicts of Interest: The authors declare no conflict of interest.

References

1. Bedogni, A.; Campisi, G.; Agrillo, A.; Fusco, V. *Expert Commission SICMF-SIPMO: Raccomandazioni Clinico-Terapeutiche Sull'osteonecrosi delle ossa Mascellari Associata a Farmaci Bisfosfonati e sua Prevenzione*; version 1.1; CLEUP: Padova, Italy, 2013.
2. Pellegrino, G.; Taraschi, V.; Vercellotti, T.; Ben-Nissan, B.; Marchetti, C. Three-Dimensional Implant Positioning with a Piezosurgery Implant Site Preparation Technique and an Intraoral Surgical Navigation System: Case Report. *Int. J. Oral Maxillofac. Implants* **2017**, *32*, e163–e165, doi:10.11607/jomi.5800.

© 2019 by the authors. Licensee MDPI, Basel, Switzerland. This article is an open access article distributed under the terms and conditions of the Creative Commons Attribution (CC BY) license (http://creativecommons.org/licenses/by/4.0/).

Extended Abstract

Ameloblastic Fibrosarcoma: Report of a New Case [†]

Andrea Tesei [1,*], Marco Mascitti [2], Filiberto Mastrangelo [3], Vera Panzarella [4], Alessandra Nori [1] and Andrea Santarelli [2]

1. Department of Surgical and Special Odontostomatology Umberto I General Hospital, Marche Polytechnic University, 60126 Ancona, Italy; alessandra.nori@ospedaliriuniti.marche.it
2. Department of Clinical Specialistic and Dental Sciences, Marche Polytechnic University, 60126 Ancona, Italy; marcomascitti86@hotmail.it (M.M.); andrea.santarelli@staff.univpm.it (A.S.)
3. Department of Clinical and Experimental Medicine, University of Foggia, 71122 Foggia, Italy; filiberto.mastrangelo@unifg.it
4. Department of Surgical, Oncological and Oral Sciences, University of Palermo (DICHIRONS), 90127 Palermo, Italy; panzarella@odonto.unipa.it
* Correspondence: andrea.tesei@outlook.it; Tel.: +39-071-2206226
† Presented at the XV National and III International Congress of the Italian Society of Oral Pathology and Medicine (SIPMO), Bari, Italy, 17–19 October 2019.

Published: 12 December 2019

Ameloblastic fibrosarcoma (AFS) is defined as a malignant mesenchymal tumor in which the epithelial component is cytologically benign and the mesenchymal component shows cytological features of malignancy [1]. AFS is the most common subtype of odontogenic sarcomas and is considered to be the malignant counterpart of fibromas of odontogenic origin [2].

AFS can occur in patients of any age, with more frequent manifestations in the third decade, predominantly male (M:F ratio of 1.5). Histologically, stromal component of AFS is characterized by malignant features of sarcoma.

Clinically, AFS grows progressively, resulting painful in many cases and causing dysesthesia or paresthesia when it involves the nerves. The non-specific symptomatology makes diagnosis difficult. AFS always appears as a radiolucent lesion with irregular margins, mainly located in the mandible. Although treatment is predominantly surgical, recurrence and metastasis rates are high.

A 55-year-old patient was reported for the evaluation of a worsening swelling from right mandibular area accompanied by pain on palpation.

Palpation revealed a non-tender, non-fluctuant and hard swelling. The neck was soft with no evidence of lymphadenopathy or tenderness. Radiographic examination was conducted using panoramic radiograph and Cone Beam Computed Tomography (CBCT), revealing a radiolucent lesion on the left mandible, closely associated with the element 4.5 and the mandibular canal.

Under general anesthesia a mucoperiosteal flap was conducted in vestibular surface of right mandibula from 4.7 area to 3.2. Using piezoelectric bone scalpel. Osteotomy was performed and the lesion were enucleated.

Histopathology showed malignant features of sarcoma, such as nuclear crowding, with hypercellularity and variable degrees of cytological atypia. On the contrary, the odontogenic epithelial component is bland and cytologically benign. The histopathological features suggested diagnosis of AFS.

A new resection, after planning on stereolithographic models, was necessary for the extension of the surgical margins and to reduce the risk of recurrence, with the colleagues of the Sant'Orsola hospital in Bologna. Post operation period was favorable for the patient.

Conflicts of Interest: The authors declared no conflicts of interest.

References

1. El-Naggar, A.; Chan, J. *WHO Classification of Head and Neck Tumours*, 4th ed.; IARC: Lyon, France, 2017; p. 214.
2. Mascitti, M.; Togni, L. Peripheral Odontogenic Myxoma: Report of Two New Cases with a Critical Review of the Literature. *Open Dent. J.* **2018**, *12*, 1079–1090, doi:10.2174/1874210601812011079.

© 2019 by the authors. Licensee MDPI, Basel, Switzerland. This article is an open access article distributed under the terms and conditions of the Creative Commons Attribution (CC BY) license (http://creativecommons.org/licenses/by/4.0/).

Extended Abstract

Unusual Manifestations of Oral Follicular Lymphoid Hyperplasia Mimicking Oral Lichen Planus [†]

Matteo Val [1,*], Margherita Gobbo [1], Marco Rossi [1], Mirko Ragazzo [1] and Luca Guarda Nardini [2]

[1] Unit of Oral and Maxillofacial Surgery, Ca Foncello Hospital, 31100 Treviso, Italy; marghe87gobbo@gmail.com (M.G.); marco.rossi@aulss2.veneto.it (M.R.); mirkoragazzo@hotmail.com (M.R.)
[2] Head Physician, Unit of Oral and Maxillofacial Surgery, Ca Foncello Hospital, 31100 Treviso, Italy; luca.guarda@unipd.it
* Correspondence: matteo.val@outlook.it; Tel.: +39-3402313851
[†] Presented at the XV National and III International Congress of the Italian Society of Oral Pathology and Medicine (SIPMO), Bari, Italy, 17–19 October 2019.

Published: 12 December 2019

Follicular lymphoid hyperplasia (FLH) of the oral cavity is a rare and poorly understood lymphoproliferative disorder which may be confused clinically and histologically with malignant lymphoma. The condition has been described in different districts: notably skin, gastrointestinal tract, lungs, nasopharynx, larynx, and breasts. Rarely, the oral cavity may be involved [1]. The disease occurs in a wide age range, namely 38 to 79 years old patients. FLH is more common in women (ratio 3:1) [2]. Clinically, the manifestation is a firm, painless, nonulcerated, slowly growing mass or swelling on the one side of the palate. Occasionally, the lesions may be multifocal, and the patients may have bilateral involvement. Usually, the lesion is soft and either colored or non-colored [3].

A 45-year-old female non-smoking patient came to our hospital complaining of relapsing-remitting pain on bilateral buccal mucosa.

She was suffering from Hashimoto's Thyroiditis, but she was not assuming any replacement therapy. She underwent bimaxillary Orthognathic Surgery several years ago.

Clinical examination revealed the presence of bilateral atrophic lesions surrounded by white striae, involving the buccal mucosa (Figure 1). The clinical suspect was of a symptomatic lichen planus. An incisional biopsy of the left buccal mucosa was performed and the pathological assessment showed hyperplastic aspect of the epithelium but otherwise unremarkable. The subepithelial tissue contained a dense follicular lymphoid infiltrate. The interfollicular tissue contained small lymphocytes, occasional large lymphocytes, plasmacells and a few eosinophils (CD20+, CD3+, Regular: Bcl2, Bcl6, CD10, Mib1) (Figure 2). An incisional biopsy was repeated on the right buccal mucosa with the same pathological assessment. The specimens were analysed by a second pathologist, who confirmed the diagnosis.

To complete the diagnostic process, in agreement with the haematologist, blood exam, an ultrasonography of the abdomen, a chest x-ray and a protein electrophoresis were performed. The exams ruled out any systemic involvement. The joint assessment of such results and of the previous investigations allowed the diagnosis of oral lymphoid hyperplasia. The patent has been followed up for 10 months with no sign of worsening of the lesions, which remained light symptomatic.

Oral manifestations of FLH, have been reported in just around 30 cases; they showed the presence of swelling in particular of the hard palate. To the best of our knowledge, this is the first reported case of bilateral buccal mucosa involvement mimicking lichen planus.

FLH is a rare and benign lymphoproliferative disorder, interdisciplinary efforts are mandatory to avoid diagnostic time delay.

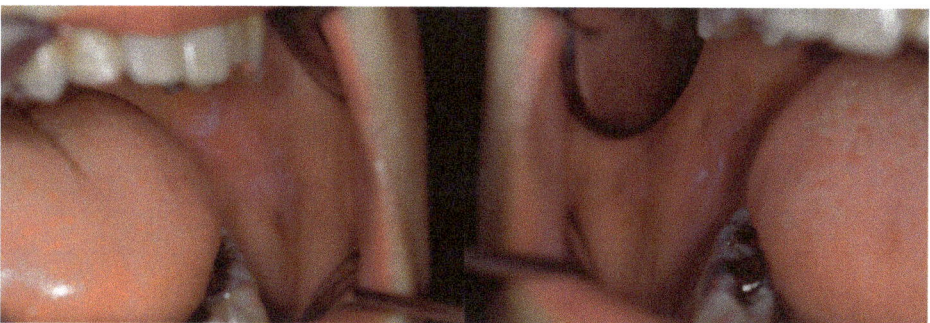

Figure 1. Clinical aspect of the FLH.

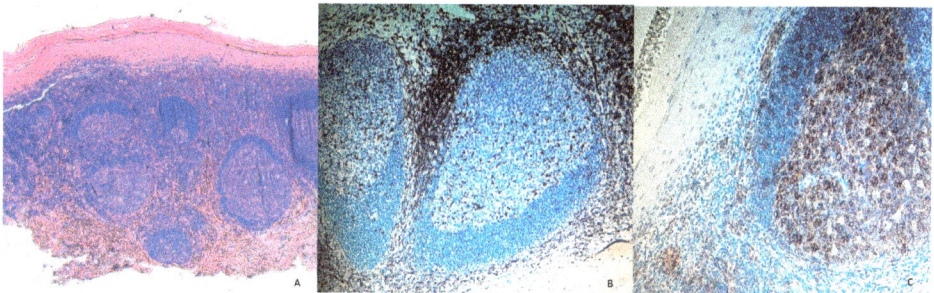

Figure 2. Pathological assessment: H&E staining (**A**); magnification ×4, positive IHC staining for CD3 (**B**); magnification ×20 and CD20 (**C**) magnification ×20.

Conflicts of Interest: The authors declared no conflict of interest.

References

1. Menasce, L.P.; Shanks, J.H.; Banerjee, S.S.; Harris, M. Follicular lymphoid hyperplasia of the hard palate and oral mucosa: Report of three cases and a review of the literature. *Histopathology* **2001**, *39*, 353–358.
2. Kolokotronis, A.; Dimitrakopoulos, I.; Asimaki, A. Follicular lymphoid hyperplasia of the palate: Report of a case and review of the literature. *Oral Surg. Oral Med. Oral Pathol. Oral Radiol. Endod.* **2003**, *96*, 172–175.
3. Jham, B.C.; Binmadi, N.O.; Scheper, M.A.; Zhao, X.F.; Koterwas, G.E.; Kashyap, A.; Levy, B.A. Follicular lymphoid hyperplasia of the palate: Case report and literature review. *J. Craniomaxillofac. Surg.* **2009**, *37*, 79–82.

 © 2019 by the authors. Licensee MDPI, Basel, Switzerland. This article is an open access article distributed under the terms and conditions of the Creative Commons Attribution (CC BY) license (http://creativecommons.org/licenses/by/4.0/).

Extended Abstract

Gene Expression Profiles in Surgical Excision Margins Detected by Tissue Auto-Fluorescence (VELscope™) in Oral Potentially Malignant Disorders (OPMDs) and Oral Squamous Cell Carcinoma (OSCC) †

Caterina Buffone [1], Flavia Biamonte [2] and Amerigo Giudice [1,*]

1. School of Dentistry, Department of Health Sciences, Magna Graecia University, 88100 Catanzaro, Italy; caterinabuffone19@gmail.com
2. Laboratory of Biochemistry and Cellular Biology, Department of Experimental and Clinical Medicine, Magna Graecia University, 88100 Catanzaro, Italy; flavia.biamonte@unicz.it
* Correspondence: a.giudice@unicz.it
† Presented at the XV National and III International Congress of the Italian Society of Oral Pathology and Medicine (SIPMO), Bari, Italy, 17–19 October 2019.

Published: 22 December 2019

1. Introduction

Oral squamous cell carcinoma (OSCC) is often preceded by "oral potentially malignant disorders" (OPMDs), "morphological alterations amongst which some may have an increased potential for malignant transformation", including leukoplakia, erythroplakia, oral submucous fibrosis, oral lichen planus, and actinic keratosis [1]. OSCC pathogenesis is a multistep process in which multiple genetic events occur that alter the normal functions of oncogenes and tumor suppressor genes [2]. Recently, tissues autofluorescence (AF) has demonstrated a better efficiency in detection of oral mucosa alterations. However, few studies providing the molecular validation of this application are available. The aim of this study was to analyze the gene expression of 45 genes using RNA sequencing evaluating the accuracy of AF imaging compared to white light for determination of bioptic margins in OPMDs and OSCC.

2. Material and Methods

The study population was represented by 7 patients with suspicion of OPMD or OSCC. After medical anamnesis, the lesions were evaluated under white light and then with VELscope ™ (Figure 1). Bioptic specimens were taken at 3 different sites for each lesion: centre (L), white light margin (MC) and AF visualized margin (MF) (Figure 2). Custom-made TaqMan arrays containing 45 were used to perform gene expression analysis. Data were analyzed by Multiple t test considering significant a p value < 0.05. Gene Ontology was used to biological pathway analysis. Analyzed genes were involved in important growth pathways such as support of cell proliferation, cell adhesion and angiogenesis [3].

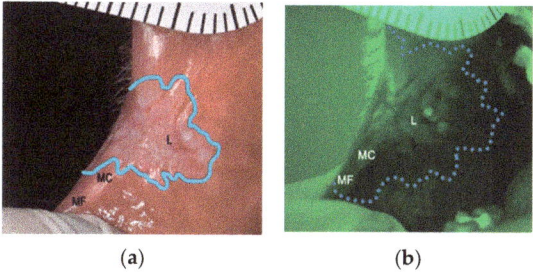

Figure 1. (**a**) Example of an OPMD (leukoplakia) on the right cheek viewed under white light; (**b**) Leukoplakia displaying loss of autofluorescence when visualised using VELscope.

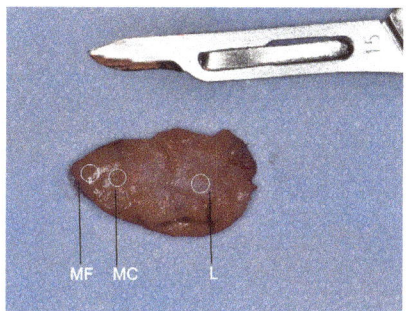

Figure 2. The regions where three biopsies were taken: at the centre of the lesion (L), clinical margin (MC), and autofluorescence margin (MF).

3. Results

The analysis was carried out on a total of 21 samples. Multiple comparisons were performed between the samples. An overview of the statistically significant DEGs across the all samples analyzed, provided in Table 1, shows a greater molecular difference between the L vs MC or L vs MF margins in comparison with MC vs MF margins, in both groups. However, the existence of DEGs between the MC and MF margins supports the importance of the molecular distinction between these two bioptic margins (Table 1). In patients with leukoplakia, up- and down-regulated genes in MC margin were involved in the mechanisms of regulation of apoptosis, cell motility, collagen metabolism and the immune response. In patients with OSCC the MC margin has showed a partially overlapping molecular profile with the L sample. The molecular divergence between MC and MF margins is the most important finding of our study.

Table 1. Number of DE genes in the whole set of samples analyzed.

	Margins	DEGs (n°)
	L vs MC	22
OSCC	L vs MF	21
	MC vs MF	10
	L vs MC	21
OPMD	L vs MF	20
	MC vs MF	9

4. Conclusions

The present study portrays a molecular dysregulation decreasing from the centre to the different detected margins of lesions. Our results suggest that a "partially transformed" cell population exists in MC and support the accuracy of AF, with respect to the white light alone, in the margin identification during the OPMDs or OSCC biopsy.

References

1. Farah, C.S.; Kordbacheh, F.; John, K.; Bennett, N.; Fox, S.A. Molecular classification of autofluorescence excision margins in oral potentially malignant disorders. *Oral Diseases* **2018**, *24*, 732–740, doi:10.1111/odi.12818.
2. Kordbacheh, F.; Bhatia, N.; Farah, C. Patterns of differentially expressed genes in oral mucosal lesions visualised under autofluorescence (VELscope™). *Oral Diseases* **2016**, *22*, 285–296, doi:10.1111/odi.12438.
3. Hanahan, D.; Weinberg, R.A. Hallmarks of Cancer: The Next Generation. *Cell* **2011**, *144*, 646–674, doi:10.1016/j.cell.2011.02.013.

© 2019 by the authors. Licensee MDPI, Basel, Switzerland. This article is an open access article distributed under the terms and conditions of the Creative Commons Attribution (CC BY) license (http://creativecommons.org/licenses/by/4.0/).

MDPI
St. Alban-Anlage 66
4052 Basel
Switzerland
Tel. +41 61 683 77 34
Fax +41 61 302 89 18
www.mdpi.com

Proceedings Editorial Office
E-mail: proceedings@mdpi.com
www.mdpi.com/journal/proceedings

www.ingramcontent.com/pod-product-compliance
Lightning Source LLC
LaVergne TN
LVHW071948080526
838202LV00064B/6702